Soil Erosion

Board of Trustees

Ernest Brooks, Jr.
 Chairman
William H. Whyte
 Vice Chairman
T. F. Bradshaw
John A. Bross
Louise B. Cullman
Gaylord Donnelley
Maitland A. Edey
Charles H. W. Foster
David M. Gates
Charles M. Grace
Philip G. Hammer
Walter E. Hoadley
William T. Lake
Richard D. Lamm
Melvin B. Lane
David Hunter McAlpin
Ruth H. Neff
Eugene P. Odum
Richard B. Ogilvie
Walter Orr Roberts
James W. Rouse
William D. Ruckelshaus
Anne P. Sidamon-Eristoff
Donald M. Stewart
George H. Taber
Henry W. Taft
Pete Wilson
Rosemary M. Young

William K. Reilly
 President

The Conservation Foundation is a nonprofit research and communications organization dedicated to encouraging human conduct to sustain and enrich life on earth. Since its founding in 1948, it has attempted to provide intellectual leadership in the cause of wise management of the earth's resources.

Soil Erosion
Crisis in America's Croplands?

Sandra S. Batie

The Conservation Foundation
Washington, D.C.

To Neal,
for it is his generation that will
inherit the earth we leave

Soil Erosion: Crisis in America's Croplands?
© 1983 by The Conservation Foundation
All rights reserved. No part of this book may be reproduced in any form without permission of The Conservation Foundation.

The photograph on page 50 was provided by the U.S. Army Corps of Engineers. All other photographs were provided by the U.S. Department of Agriculture, Soil Conservation Service.

Cover design by Graphics in General, Washington, D.C.
Typography and composition by VIP Systems, Inc., Alexandria, Virginia
Printed by Todd/Allan Printing Company, Washington, D.C.

The Conservation Foundation
1717 Massachusetts Avenue, N.W.
Washington, D.C. 20036

Library of Congress Cataloging in Publication Data

Batie, Sandra S.
  Soil Erosion.

  Includes bibliographical references.
  1. Soil erosion—United States. 2. Soil conservation—United States. 3. Soil conservation—Government policy—United States. I. Conservation Foundation. II. Title.
  S624.A1B33   1983   631.4'5'0973   83-1942
  ISBN 0-89164-068-1

Contents

Foreword by William K. Reilly   vii
Preface   ix
Executive Summary   xiii
Chapter 1. Renewed Concern over Soil Erosion   1
   Early Conservation Efforts   1
   The Expanding Export Market   5
   The Environmental Movement   7
   The Accountability Crunch   8
   The Magnitude of the Problem   10
Chapter 2. The Nature and Extent of Soil Erosion   13
   The Properties of Soil   13
   The Physical Process of Erosion   14
   Factors Influencing Erosion   18
   Measuring Erosion Losses   26
Chapter 3. The Effects of Erosion   37
   Yields and Erosion   37
   Environmental Impacts   44
Chapter 4. Techniques for Reducing Soil Erosion   55
   Changing the Characteristics of Soil and Topography 55
   Choosing Where to Plant   55
   Crop Rotation   56
   Strip Cropping   57
   Retaining Crop Residues   57
   Reducing Wind Erosion   59
   Construction Measures   59
   Conservation Tillage Practices   60
   Conservation Tillage Trends   69
Chapter 5. Factors Affecting Farmers' Adoption
    of Conservation Practices   73
   Personal Preferences   74
   Cost of Conservation   75
   Land as an Investment   80
   Tenure Arrangements   82
   Tax Policies   83
   Loan Policies   85
Chapter 6. Present Soil Conservation Programs   89
   Major Federal Conservation Programs   92
   Assessing the Programs   93

Policy Changes	96
Water Quality Programs	97
Commodity Programs	99
Local Institutions	101
State Programs	103
Chapter 7. Strategies for Encouraging Soil Conservation	**109**
Strategy Choices	112
The Shape of Conservation Policy	123
Appendix. Measuring Soil Erosion Losses	**129**
Index	**133**

Foreword

Soil erosion is not yet a crisis in this country. The products of American farms still feed a good part of the world, and will continue to do so for some time. But despite 50 years of federal efforts and billions of dollars expended to conserve productive soils, erosion persists as one of this country's major conservation problems. Erosion not only robs farmland of its fertility, it also seriously pollutes the nation's waterways. It may even have accelerated as farmers responded during the past decade to economic opportunities and pressures by planting more cropland, including marginally productive lands and lands prone to erosion, and abandoning conservation measures. Ironically, most Americans believe our soil erosion problem was resolved during the 1930s when severe droughts and dust storms swept across the prairies and midwestern soil accumulated on windowsills of the Capitol in Washington, D.C.

For this generation, with its new farm technology and its dim recollection of the social and economic catastrophe to which soil erosion contributed in the 1930s, we need to put this issue on the nation's agenda. This book is part of the effort to do that. If Americans do not take seriously the accumulating evidence about the extent and consequences of erosion, the country's agricultural future may be undermined, perhaps not this decade or next, but sometime early in the twenty-first century. If this happens, then I believe decision makers of today would have violated a trust they hold for future generations of Americans.

Curbing the loss of fertile soils so that the land provides adequately for future Americans is a concern that major conservation groups in recent years have not seriously addressed, although conservationists have a critical role in articulating and explaining the problem, and its consequences, and in ensuring serious consideration of responses.

Fortunately, as Sandra Batie discusses, reasonable solutions to the soil erosion problem do exist. Unfortunately, the economic plight of American farmers in 1983 leaves precious few resources to allocate to erosion control. Progress will require not only the understanding and cooperation of farmers but will also require that they reduce their debt and increase their income. The education and cooperation of the non-farm population, legislators, and government officials will also be necessary to the development of effective regional and national soil conservation policies. We hope Sandra Batie's book will help inform and educate lay readers toward this end.

Soil Erosion follows in the tradition of earlier Conservation Foundation efforts to document and publicize the nature and consequences of natural resource problems. After Fairfield Osborn, Jr., and his associates established The Conservation Foundation in 1948, the first project they undertook was a cooperative venture with the Food and Agriculture Organisation of the United Nations, the Pan American Union, and others to survey the incidence of soil erosion in the Western Hemisphere, focusing on developing countries in Central and South America. In some places, this represented the first time that data on soils, terrain, weather, and other factors affecting erosion were gathered and analyzed. While the effort was severely handicapped by lack of information, its choice as an initial project symbolized the importance the founder of The Conservation Foundation attached to the maintenance of soil productivity. The problem looms large in our own time, in our own country, and thus we return to it.

We are deeply grateful to The Ford Foundation, The Joyce Foundation, the Rockefeller Brothers Fund, and The Rockefeller Foundation for their support of Dr. Batie's research and publication.

> William K. Reilly
> President
> The Conservation Foundation

Preface

When I first became interested in soil conservation—around 1979—it was from the perspective of someone who wanted to learn more. I thought, at the time, that there must be numerous books addressing soil conservation issues that would answer my questions. After a search of available literature, I found that most works could be placed in two general categories. The first category was what I term "awareness" books—books that hoped to provoke the reader into taking action to "stop erosion." This approach depended on emotional descriptions of devasted societies of the past, whose ruin was assigned to lack of soil stewardship, or imagistic scenarios of future famines when America's soil was gone.

The other category of books was mainly technical. These contained detailed descriptions of how erosion occurred, how to use the soil-loss prediction equation, and how to reduce erosion through such techniques as terrace or farm pond construction.

There were very few books or articles that provided what I felt was good policy analysis of America's soil erosion problems. I wanted to know: What are the scope and nature of the erosion problem? Why is it a problem, particularly after years of conservation programs? Would technologies, such as improved plant strains and fertilizers, substitute for any erosion damage? And, if so, at what cost? Are farmers adopting those conservation measures deemed worthwhile? If not, why not? What is the justification for public policies addressing soil conservation? Unfortunately, the few books and articles that tried to provide careful analysis of these types of questions were severely handicapped by lack of data.

As I was investigating soil erosion issues in the United States, new data were becoming available that would, ultimately, provide some insights into my questions. The information resulted mainly from the Soil Conservation Service's 1977 National Resource Inventory and the U.S. Department of Agriculture's response to the Soil and Water Resources Conservation Act of 1977. The availability of the data and increased interest in soil conservation also meant that many agronomists, agricultural economists, political scientists, sociologists, and other researchers were investigating soil conservation issues in a scientific manner. While there were still many knowledge gaps, it was now possible to begin to analyze the effectiveness of past programs, to determine the location and severity of the nation's erosion problems, and to determine the factors that influence farmers' conservation behavior.

As I became familiar with the information available, I was convinced that soil conservation was an important national issue and that the development of an improved soil conservation policy was a matter worthy of the nation's attention. It was then that the idea of publishing a book seemed appropriate since, in a pluralistic society, good policy is most likely to emerge if concerned citizens—farmers and environmentalists, ranchers and state legislators, rural and urban people—are kept abreast of available and accurate information.

With these citizens as my perceived audience, I have written a book that evaluates the (incomplete) information on soil erosion in the United States. I have attempted to address the limitations of the information available, yet at the same time to draw what implications I can for the structuring of an improved public policy for soil conservation.

Because this book offers policy analysis, I hope it also proves useful to university undergraduate and graduate students exploring soil conservation policy issues as well as to soil conservation district and university extension personnel who need such information to develop educational programs.

Many people have contributed to this work and made it better than it would have been without their help. I was assisted through much of the research by two very capable interns: Carol Zabin and Tamara Vance. I like to think that their research experience at The Conservation Foundation was at least partly responsible for their decisions to pursue advanced degrees in economics and agricultural economics.

Many people assisted me in my investigation of actual erosion problems in Tennessee, Iowa, New Mexico, and Delaware. Wintfred Smith, a limnologist at the University of Tennessee at Martin, was extremely helpful in discussing the Reelfoot Lake soil erosion problems as were Ralph Burruss, Reelfoot Lake State Park superintendent, and his assistant, Jimmy Cox. Pat Barnett, John

PREFACE xi

Butler, Bob Fisher, Arthur Fuqua, Estel Hudson, Larry Kimery, Joseph Martin, and Doyle Tucker all helped me gain a better perspective of the soil erosion problem in west Tennessee. Loren Brandsma, Ted Hall, Neil Hamilton, William Hawks, Merle Lawyer, Bud Lyons, Liza Ouram, John Timmons, and Lawrence Vance were similarily helpful on my visits to Iowa. Lonni Ashcraft and Dick Shaw provided background information on some of the erosion problems associated with the more arid regions of New Mexico. Corin Carty provided information on Delaware's conservation programs.

Throughout the book I have used anonymous quotations. Some, but not all, of these interviews were from my discussions with the people listed above. In all cases, however, the interviews were actual events and the thoughts expressed were those of the person interviewed.

Many colleagues aided my research, either by providing background research discussions or through helpful review comments. These people include Frank Bell, Norm Berg, Gene Carson, Toby Clark, Pierre Crosson, Terry Davies, Mary Gardner, Mack Gray, Robert Healy, Beatrice Holmes, Randy Kramer, Larry Libby, Jake Looney, Bill Park, Wayne Rasmussen, Barbara Rodes, Dan Taylor, and Neil Sampson. I have also come to appreciate the role of an editor in producing any book. Special thanks are due to Beth Davis, Jo Tunstall, and Bethany Brown for performing this function at various stages of this book's evolution. Tony Brown worked through many nights typing and retyping the manuscript.

Finally, I owe a special debt of gratitude to all the staff members who give The Conservation Foundation its reputation of a pleasant place in which to work. My year with The Conservation Foundation will always remain a special memory.

Sandra S. Batie
Blacksburg, Virginia

Executive Summary

After nearly a half-century of government programs to combat erosion, cropland erosion is still occurring at an estimated rate of 2 billion tons of soil per year. Some critics argue that soil erosion is worse than it was before the programs began. Some have concluded that America's croplands are deteriorating rapidly, making the situation a crisis. Others, pointing to the nation's history of bountiful harvests, argue that while problems caused by erosion exist—water pollution from land runoff for one—there is no crisis and little reason to commit additional public resources to the problem.

Who is right? *Soil Erosion: Crisis in America's Croplands?* poses this question and finds answers in analysis of several dimensions of the erosion problem.

Soil erosion—whether caused by wind, rain, or snowmelt—reduces the depth of topsoil, impairs soil's capacity to supply water, and thwarts infiltration of water and air into the soil. The severity of erosion is the product of many factors, including climate, soil type, topography, and farming practices.

Although measurement of soil loss is difficult, widely used tools show that 4 regions have severe water-caused erosion problems:
- the Palouse and Nez Perce areas and the Columbia Plateau of western Idaho, eastern Washington, and north-central Oregon;
- the wind-deposited soils of Nebraska, Kansas, Iowa, and Missouri;
- the uplands of the southern Mississippi Valley; and
- the cultivated areas in Aroostook County, Maine.

Almost 70 percent of the erosion exceeding 5 tons per acre occurs on less than 8.6 percent of the total cropland acreage.

Measuring crop yield reductions caused by soil erosion is difficult because, over time, farmers may substitute increasing amounts of fertilizers or use other technologies to enhance soil's natural fertility. Also, the effects of erosion on crop yields differ by soil type, by crop, and by measurement practices. Studies indicate, however, that a relationship exists between soil erosion and reduced crop yields on many soils. If erosion has reduced the water-holding capacity of the soil, the rooting depth available to the plant, or the water infiltration rate, adding fertilizer may not offset the yield-reducing effects of erosion.

Many experts express optimism about technological breakthroughs to reduce soil-loss problems. Other experts fear that technologically induced improvements in crop yields will not occur as quickly as in the past because of inadequate funds for agricultural research.

Soil erosion also affects air and water quality. Agriculture is blamed for most nonpoint source water pollution. Sediment in water runoff carries along fertilizer residues, pesticides, dissolved minerals (such as salts), and animal wastes (with associated bacteria). Excessive sedimentation clogs navigation channels and dirties drinking water, adding costs to rectify both, and can reduce recreational opportunities. In many areas of the country, sediment is rapidly filling inland lakes and reservoirs.

Farmers can choose among numerous management practices to reduce erosion, although all techniques are not suitable on all lands. Farmers can change (in some circumstances) the characteristics of soil and topography, choose appropriate crops, rotate crops, construct terraces or waterways, or use conservation tillage methods to reduce soil erosion.

Erosion statistics reveal, however, that many farmers are not protecting their soils. Some farmers do not recognize they have soil erosion problems. Some prefer to manage their farms in ways that exclude certain conservation practices.

Furthermore, many conservation practices do not pay for themselves. The business-minded farmer, who must remain competitive to stay in farming, is not interested in unprofitable conservation practices. Even a farmer with a strong land ethic may be financially unable to practice conservation. On many lands, however, conservation tillage creates profits.

The varying motives for landownership complicate decisions about soil conservation. Cropland is an inflation hedge for some investors; these owners may have more concern for the land's contribution to their net-worth statement than for its future production capability. Evidence suggests that these owners have not been penalized for erosion by a reduced selling price per acre when land is resold. Other factors influencing the adoption of soil conservation practices include insecure property tenure (leasing of land), tax policies, and loan policies.

A number of existing programs address soil conservation. Three major ones are the Conservation Operations Program, the Great Plains Conservation Program, and the Agricultural Conservation Program. These three have received the most criticism for failing to achieve conservation goals. Others also affect farmers' soil conservation decisions. Price support programs, for example, encourage farmers to specialize and use intensive farming practices to maximize grain production. These programs may also penalize the conservation-oriented farmer.

Most soil conservation programs are administered through local soil conservation districts (SCDs). While many SCDs have powers to regulate conservation, almost all rely on voluntary cooperation and provide farmers with cost-sharing funds to assist them. Some SCDs and states have experimented with mandatory aproaches. Since any conservation program must ultimately influence farmers, an understanding of the various factors affecting the adoption of conservation behavior is crucial for the design of an effective public conservation program.

There are numerous soil conservation strategies from which to choose. Some strategies have more potential than others for reducing erosion problems in a cost-effective manner. They include targeting of conservation efforts; removing the most erodible lands from crop production; encouraging some farmers to use low-cost practices such as reduced tillage, residue retention, and contour plowing; and using some cross-compliance strategies. Research designed to improve yields per acre and remove obstacles to soil conservation also is promising.

Alternating contour strips of corn and alfalfa reduce soil erosion on this farm in northwestern Illinois. Grass waterways are maintained in natural water courses.

Chapter 1

Renewed Concern over Soil Erosion

Jim O'Brien placed his tractor in neutral and studied the freshly plowed hillside. It was a spring day. The soil was warm enough to begin planting the year's corn crop. Indeed, the weather conditions seemed ideal. Just the right amount of moisture was left in the Iowa soil from the winter snows and early spring rains. But the farmer was not thinking about weather. He was studying the hillside. There was no mistaking it. The color of the soil was different from what it had been when he bought the farm. It was no longer the rich dark black so characteristic of the area; it was a brownish yellow. There was no mistaking the meaning of that color either. The topsoil was gone from the hillside, and only the less productive subsoil remained. O'Brien's farm was eroding, apparently severely in places, and he was concerned.

This Iowa farmer is not alone in his concern. Thousands of farmers across the country are worrying about increasing soil erosion on their lands. Millions of cropland acres are losing soil to wind and rain at rates far exceeding the natural formation of soils. The problem is widespread enough for articles to appear in newspapers and magazines under such headlines as "A Renewed Threat of Soil Erosion: It's Worse Than the Dustbowl," "Are American Farmers Exporting Their Soil?" and "The Disappearing Land." We have entered an era of renewed concern over soil erosion, and justifiably so.

Early Conservation Efforts

Protecting soil fertility has always been of some concern, even in colonial America. Many settlers, for example, understood the benefits of planting red clover in rotation with other crops, a practice that rebuilt soil fertility. Thomas Jefferson was an eloquent conservation spokesman. He recommended contour plowing, clover planting, and crop rotation to retain soil and soil fertility.

Still, such conservation efforts were adopted by few. Land was so plentiful and labor so scarce that it made little economic sense to practice much conservation. Edmund Ruffin's soil-chemistry experiments in the 1840s on his Virginia farm were dubbed Ruffin's Folly by his neighbors.[1] They considered spending time and dollars to conserve soil foolish since they were perched on the edge of a rich and underused continent. In addition, most had no reason to believe that a soil's productivity could be exhausted.

Throughout the nineteenth century, westward expansion and the Civil War meant little attention was given to soil conservation. After the war, the pressures for debt repayment led farmers to increase production. The results were both low prices for farm products and soil exploitation. If farms "wore out," they were abandoned, and farmers migrated to new lands in the West.[2]

By the 1890s, the practice of farm exploitation and abandonment had become a concern of the newly formed U.S. Department of Agriculture (USDA). One of the first USDA bulletins for farmers deplored the thousands of acres of eroded cropland abandoned each year. This bulletin, *Wasted Soils: How to Prevent and Reclaim Them*, published in 1894, urged farmers to conserve the land that they had.[3]

Despite such efforts, soil conservation did not gain the attention of a broader public until the 1920s. It did so then largely through the efforts of one man—Hugh Hammond Bennett.

Bennett became convinced of the dangers of soil erosion as early as 1903, when he was mapping soil types for the USDA's Bureau of Chemistry and Soils. While mapping, he came across example after example of soil erosion throughout the country. His experience motivated him to lecture extensively about soil erosion problems, and in 1928 he coauthored, with W. R. Chapline of the U.S. Forest Service, a report entitled *Soil Erosion: A National Menace*.

Bennett's was a messianic campaign to awaken farmers and legislators to the dangers of erosion. However, convincing legislators of the importance of a federal role in preventing soil erosion was not easy. Many thought that if landowners understood the seriousness of erosion, they would conserve without public assistance since it was in their long-term interest to do so. The first federal action, therefore, was to provide funds to be

Hugh Hammond Bennett in an eroded field.

used only for soil erosion research. Congressman James P. Buchanan of Texas, ranking minority member of the Agriculture Appropriations Committee, added an amendment to the 1933 Agriculture Appropriations Act (P.L. 70-769) that provided the secretary of agriculture with $160,000 to be used for such research. Bennett was placed in charge of the program, and he established 10 erosion-control experiment stations throughout the country.

Stronger legislation soon followed, as a result of two major events. The first was the Great Depression; the second, the terrible droughts of the early 1930s.

The Great Depression put nearly one-quarter of the labor force out of work at a time when there were no public welfare programs, unemployment insurance, or food stamps. The Depression also meant that farmers, faced with high debts and low prices, could not afford to practice soil conservation.

Therefore, in 1933, President Franklin D. Roosevelt requested Congress to authorize $5 million for soil conservation projects, both to combat erosion and to increase employment. Congress did so, under authority of the National Recovery Act of June 16, 1933 (P.L. 73-67). The money was administered by the newly created Soil Erosion Service, a temporary agency headed by Hugh Hammond Bennett in the U.S. Department of the Interior. Bennett used the money to develop demonstration projects to illustrate good soil conservation practices. The Civilian Conservation Corps and Works Progress Administration provided the labor.[4]

Then, in 1934, severe droughts hit. Dust storms carried soil from the plowed fields of the Great Plains all the way to the Atlantic Ocean. Although likely apocryphal, it is said that one of these storms rained dust on Washington, D.C., in April 1935, as Congress was considering H.R. 7054, the Soil Conservation Act. The act, which later passed, declared the United States was:

> to provide permanently for the control and prevention of soil erosion and, thereby, to preserve natural resources, control floods, prevent impairment of reservoirs, and maintain the navigability of rivers and harbors, protect public health, public lands, and relieve unemployment . . .[5]

The act also established the Soil Conservation Service (SCS) as a permanent agency within the Department of Agriculture.

The Soil Conservation Act was followed by passage of the Soil Conservation and Domestic Allotment Act of 1936, which provided government payments for adoption of soil conservation practices, primarily to reduce acreages of "soil-depleting" crops.

In 1937, the Standard State and Soil Conservation District Model Law was formulated with hopes that it would be adopted (in some form) by each state. Twenty-two states passed such a law in 1937. By 1947, all states had passed soil conservation district enabling laws. The model law outlined a process by which local soil conservation districts could be voluntarily established within a state and used to promote soil conservation practices. Today there are 2,950 soil conservation districts, which collectively cover approximately 2.2 billion acres.[6]

Although various programs were added to those of the 1930s over the next three decades, soil erosion itself aroused little public interest. To most Americans, it seemed that action had been taken to deal effectively with the problem. With the events of the 1970s, all this changed.

The Expanding Export Market

The 1970s saw the first major interruption of several decades of U.S. agricultural surpluses.

Before 1973, American agriculture had excess resources, both in land and in the number of farmers. These resources, coupled with government programs that supported some farm prices above market levels, resulted in substantial annual surpluses of agricultural commodities. In the face of overflowing silos, soil conservation was perceived as an issue of little urgency.

But grain exports increased dramatically in 1973, when the Soviet Union began buying large quantities of foreign grain. Grain exports in 1973 were nearly double those of 1972. The 1974 prices of wheat, soybeans, and corn rose 208 percent, 133 percent, and 128 percent, respectively, above 1970 prices. The increase in harvested cropland was spectacular and predictable: from 1973 to 1974 there was a net increase of 24 million harvested acres.

With agricultural exports considered essential to improving the nation's balance of payments, farmers of the 1970s were encouraged by Secretary of Agriculture Earl Butz to plant "fence row to fence row." And they did. Harvested lands were used more intensively. Pastures were plowed and planted. Marginal agricultural lands were cultivated, often for the first time. From 1967 to 1977, over 2 million acres of newly cultivated cropland came from lands with poor soil.

As farmers plowed under more land to take advantage of rising

Grain being loaded for export from the port of New Orleans.

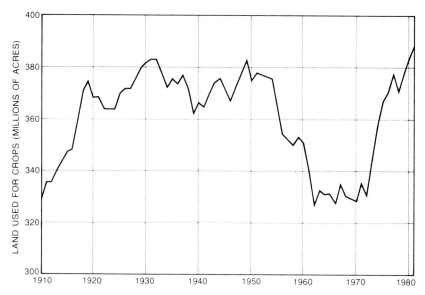

Figure 1.1. Cropland in actual production in the United States.
SOURCE: S. S. Batie and R. G. Healy, "The Future of American Agriculture," *Scientific American* 248(2):45-53 (1983).

prices, old conservation practices were lost. Narrow terraces, suitable for the smaller and lighter farm machinery of the 1960s, were plowed out, and contouring of fields was abandoned because of incompatibility with the machinery of the 1970s.

By 1981, in response to the increased export demands, harvested cropland reached 391 million acres, up substantially from less than 335 million acres in 1972 (see figure 1.1). Partly as a result of these trends, the quality and quantity of America's croplands became a national concern.

The Environmental Movement

The environmental movement that gained visibility in the 1970s also drew attention to soil erosion. With water quality a major concern of environmentalists, it was only a matter of time before the demands for cleaner water focused on one of the chief pollutants: soil eroded from agricultural lands. It has been estimated that over 33 percent of the oxygen-demanding loads, 66 percent of the phosphorus, and 75 percent of the nitrogen discharged into streams come from dispersed agricultural sources.[7]

The environmental concerns introduced a new set of players to soil conservation issues, as USDA officials soon discovered.

> We learned the unique blend of federal, state, and local participation employed in the soil conservation movement had been a weakness as well as a strength. ... That is, USDA agencies and conservation districts thought they had involved the "right" leaders and the "right" groups who represented the whole community and its aims and concerns.
>
> That was not always the case. People whose interest we did not know about showed up in meeting rooms or courtrooms. Objectives and practices that we assumed had consensus suddenly were being questioned.[8]

Although the SCS and other federal agencies had programs that dealt indirectly with water pollution caused by erosion, the "new players" encouraged a more direct federal role through the Federal Water Pollution Control Act (FWPCA, now the Clean Water Act) and its amendments.

There is no provision in the FWPCA nor in its amendments for direct federal regulation of nonpoint pollution,* but there is an objective to make the nation's waters "fishable and swimmable" by 1983. To meet this objective, the amendments state that "areawide waste treatment management planning processes be developed and implemented to ensure adequate control of [all] sources of pollutants in each state."[9]

In Section 208, the FWPCA specifically states that such areawide planning should include significant nonpoint sources of pollution and should identify methods to control such pollution. It is this provision that has provided an impetus for water quality officials, conservation district officials, and USDA officials to coordinate efforts to manage nonpoint water pollution. These efforts include the experimental USDA Rural Clean Water Program, which has as its objective obtaining long-term water quality goals through contractual agreements with rural landowners.

The Accountability Crunch

In the late 1970s, serious criticisms of the nation's soil conservation programs began to emerge. After almost 50 years of public

---

*Nonpoint pollution is pollution from sources that cannot be pinpointed; rather, it emanates from diffuse sources. Runoff from agricultural lands is one type of nonpoint pollution.

RENEWED CONCERN OVER SOIL EROSION

Sediment pollution from the drainage area of the Loosahatchie River entering the Mississippi River 1 mile north of Memphis, Tennessee, April 1968.

programs addressing soil erosion, the programs were accused of being ineffective and expensive. Some critics even charged that soil erosion was worse than it had been before the programs were initiated.[10]

In 1977, the comptroller general of the United States prepared a General Accounting Office (GAO) report for the Congress entitled *To Protect Tomorrow's Food Supply, Soil Conservation Needs Priority Attention*. The GAO visited 283 farms throughout the nation and found that 84 percent of the farms were suffering annual soil losses above levels thought to be allowable for sustained productivity.[11]

These criticisms were accompanied by threats to cut the budget of USDA and particularly of the SCS. In late 1976, the Department of Agriculture was asked by the Senate to produce evidence that conservation programs were having an impact and should be supported. While USDA was responding to this oversight request, Congress passed legislation requiring the department to set up a formal process to evaluate soil conservation goals and methods. The Soil and Water Resource Conservation Act (P.L. 95-192), commonly known as the RCA, requires that the Department of Agriculture (1) appraise on a continuing basis the soil, water, and related resources on nonfederal land; (2) develop programs for furthering the conservation, protection, and enhancement of these resources; and (3) evaluate, annually, program performance in achieving conservation objectives.

The Magnitude of the Problem

By 1981, the USDA responded to the first of the RCA's requirements by publishing a detailed appraisal of the nation's soil and water resources.[12] The appraisal drew heavily on the 1977 National Resources Inventory, conducted by the SCS, and on the Second National Water Assessment, prepared by the Water Resources Council. The appraisal provided the best data yet available on the physical dimensions of soil erosion in the United States. It estimated that national soil erosion losses from croplands totaled 2 billion tons a year.* The sheer magnitude of the losses has caused some to question whether we are faced with an emerging

---

*A ton of soil is roughly equivalent to a cubic yard. One inch of topsoil covering 1 acre would weigh about 165 tons.

crisis in America's croplands.

Some authorities interpret the erosion estimates as alarming and ultimately life-threatening:

> [T]he erosion of soil ... adversely affects food prospects. Environmental stresses that affect the food system threaten to undermine our contemporary global civilization as they did earlier local ones. In the absence of immediate attention to [this threat] ... the struggle to make it from one harvest to the next may become a global preoccupation.[13]

At the same time, however, there has been a recent return to large surpluses of grains and soybeans. To some, it seems silly to use resources to protect soil when the historical problem has usually been the disposal of America's overly bountiful harvests. Furthermore, farmers have consistently been able to use land-saving technologies (agricultural chemicals, improved plant strains, and new farm machinery, for example) to reduce the importance of land in production.

> The key idea is that land is man-made, just as are other inputs to farm production.... With the development of modern farming methods, the output of a piece of land depends increasingly little on its natural endowments.[14]

It is easy to view the alarming statistics with a healthy skepticism. What is the truth concerning soil erosion? Is future food production in jeopardy? Are we exporting our soil as well as our food products? Are we impairing forever the health of our lakes and rivers? In short, how much soil erosion is too much?

An auxiliary but equally important question is why, after 40 years of conservation programs, is the erosion problem still of such magnitude? Have so few changes been made since Bennett's time? Could the data on the nature and extent of erosion be misleading? And, if the data are accurate, why are farmers failing to protect their own lands? Do they lack the knowledge ... the money ... the tools ... the incentive? The answers are not self-evident, but the development of appropriate public responses to the nation's cropland erosion problem depends on finding them.

References

1. D. H. Simms, *The Soil Conservation Service* (New York, N.Y.: Praeger Publishers, 1970), p. 5.

2. W. D. Rasmussen, "History of Soil Conservation, Institutions and Incentives," in H. G. Halcrow, E. O. Heady, and M. L. Cotner, eds., *Soil Conservation Policies, Institutions and Incentives* (Ankeny, Iowa: Soil Con-

servation Society of America, 1982), p. 3-10.

3. *Ibid.*, p. 5.

4. R. Dallavalle and L. V. Mayer, *Soil Conservation in the United States: The Federal Role, Origins, Evolution and Current Status*, CRS Report no. 80-144S (Washington, D.C.: Congressional Research Service, September 1980), p. CRS-3.

5. P.L. 74-46, 49, Stat. 163 (1935).

6. Dallavalle and Mayer, *Soil Conservation in the United States*, p. CRS-9.

7. U.S. Water Resources Council, *Second National Water Assessment, The Nation's Water Resources, 1975-2000*, vol. 1, summary (Washington, D.C.: U.S. Government Printing Office, 1978), p. 60.

8. D. G. Unger, "Evolution of Institutional Arrangements: A Federal View," in *Soil Conservation Policies: An Assessment* (Ankeny, Iowa: Soil Conservation Society of America, 1979), p. 30.

9. 33 U.S.C. Section 101(a)(5); 33 U.S.C. Section 1251(a)(5) as referenced by B. H. Holmes, *Institutional Bases for Control of Nonpoint Source Pollution under the Clean Water Act with Emphasis on Agricultural Nonpoint Sources* (Washington, D.C.: U.S. Environmental Protection Agency, November 1979), p. 3.

10. T. Barlowe, "Three-Quarters of the Conservation Job Not Being Done," in *Soil Conservation Policies: An Assessment* (Ankeny, Iowa: Soil Conservation Society of America, 1979), p. 128-32.

11. U.S. Comptroller General, *To Protect Tomorrow's Food Supply, Soil Conservation Needs Priority Attention*, CED 77-30 (Washington, D.C.: General Accounting Office, 1977), p. i.

12. U.S. Department of Agriculture, *Soil and Water Resources Conservation Act, 1980 Appraisal Part I, Soil, Water, and Related Resources in the United States: Status, Conditions and Trends* (Review Draft) and *Appraisal Part II, Soil, Water, and Related Resources in the United States, Analysis of Resource Trends* (Review Draft) (Washington, D.C.: U.S. Department of Agriculture, 1981).

13. L. R. Brown, *Building a Sustainable Society* (New York, N.Y.: W. W. Norton & Co., 1981), p. 6.

14. J. L. Simon, "Are We Losing Ground?" *Illinois Business Review* 37(3):3 (1980).

Chapter 2

The Nature and Extent of Soil Erosion

An understanding of the effects of soil erosion requires understanding the various functions performed by soil. Soil erosion affects crops by reducing the depth of the topsoil and thereby affects availability of soil nutrients, the capacity of the soil to supply water, and the infiltration of water and air into the soil. The severity of these effects differs widely by region, by soil, by farming practice, by type of crop, and by growing season.

The Properties of Soil

Soils, and the plants and microorganisms that grow in soils, form an ecosystem in which the four components necessary for crop growth are recycled. These four components are mineral particles, organic materials, water, and air. The productivity of a soil depends on the proportions and interactions of the four components.

Mineral particles come directly from rock material that is weathered by physical and chemical processes and is further altered by microbial action. The size and amounts of these particles determine the soil's texture. For example, clays are composed of small particles; silts and sands consist of larger particles; and loam is a combination of sand, clay, and silt particles.

Organic compounds are derived mainly from plant and animal residues, which are broken down by microorganisms. Organic matter greatly affects soil structure by forming large pore spaces in which air and water can combine to support plant growth. Organic matter is also a primary source of plant nutrients, especially nitrogen. Furthermore, porosity and the capacity to retain moisture, key factors in determining a soil's productivity, tend to increase with increases in organic matter.

Since soil nutrients can be artificially supplemented by using fertilizers, the capacity of a soil to supply moisture is frequently the most important determinant of crop yields. Water-holding capacity depends on soil texture and on the depth of the soil that is favorable to root development. Water-holding capacity thus can be considerably influenced by erosion. Medium-textured soils have the highest ability to supply water; coarse-textured soils have the lowest.

Soils are generally formed in distinct layers, which may range from under an inch to many feet in depth. Topsoils (called the "A horizon") generally contain a greater proportion of organic matter and soil microorganisms than subsoils. It is this uppermost layer that is the major zone of root development; it holds a large portion of the water and nutrients taken up by plants.[1]

Subsoils generally contain less organic matter, fewer nutrients, and more gravel and stones than topsoils. Frequently, the clay content of subsoils is greater than that of topsoils. More clay and less organic matter reduce water-holding capacity. If the subsoil is near the surface and inferior to the topsoil, it can limit a plant's root zone and thereby reduce plant growth.

Subsoils also are often less responsive to applied chemicals. Clay-sized iron or aluminum compounds in the soil, for example, reduce the effectiveness of applied phosphorus.[2] The effectiveness of some herbicides is reduced by low organic matter content because the binding site for the chemical herbicides is on particles of organic matter and clay.[3] Other subsoil characteristics that can inhibit plant growth are the existence of fragipans,* permanently high water tables, or chemically toxic zones in the soil. Therefore, erosion that removes topsoils to the point that plant roots reach subsoils or where subsoils are mixed throughout the plow layer can result in reduced yields and/or increased production costs.

The Physical Process of Erosion

In the spring of 1905 in Louisa County, Virginia, Hugh Hammond Bennett and his friend Bill McLendon observed a remarkable phenomenon. Bennett later wrote:

---

*Fragipans are dense and brittle layers of soil that owe their hardness mainly to extreme density or compactness rather than to high clay content or cementation. The material is so dense that roots cannot penetrate and water moves through it very slowly.

THE NATURE AND EXTENT OF SOIL EROSION

> We were stirring through the woods down there in middle Virginia when we noticed two pieces of land, side by side but sharply different in their soil quality. The slope of both areas was the same. There was indisputable evidence the two pieces had been identical in soil makeup.
>
> But the soil of one piece was mellow, loamy, and moist enough even in dry weather to dig into with our bare hands. We noticed this area was wooded, well covered with forest litter, and had never been cultivated.
>
> The other area, right beside it, was clay, hard and almost like rock in dry weather. It had been cropped a long time.
>
> We figured both areas had been the same originally and that the clay of the cultivated area could have reached the surface only through the process of rainwash—that is, the gradual removal, with every heavy rain, of a thin sheet of topsoil. It was just so much muddy water running off the land after rains. And, by contrast, we noted the almost perfect protection nature provided against erosion with her dense cover of forest.[4]

Erosion is a natural process. When lands are covered by vegetation, the rate of erosion is slow, approximately 1 inch every 100 to 250 years, and is offset by the creation of new soil. But on lands devoid of vegetation, as Bennett and his friend saw, erosion rates increase by magnitudes.

Erosion can be caused by wind or water. Water-induced erosion is of three types: sheet, rill, or gully.

Sheet erosion occurs when water removes relatively thin layers of soil from the land. It is nearly invisible to the casual observer. A farm losing 40 tons of soil per acre per year loses approximately 4 inches of soil every 15 years. That is an average loss of about one-quarter of an inch annually, not an easily discernible amount.

Rill erosion, in contrast, is quite visible. Rill erosion occurs when water from rain or melting snow, along with dislodged soil particles, concentrates in streamlets. As the runoff increases in velocity, it carries along more and more soil. Small channels, called rills, form. They look like little incisions on the face of the land, but they are easily removed with plowing.

Gully erosion is a severe form of rill erosion. It can be an exceptional nuisance and can scar a farmer's income as well as his land.* Gully erosion, too, is a product of runoff from rain-

---

*Many of America's farm owners and operators are women. However, this book uses masculine personal pronouns to refer to farmers, avoiding the

Damage from sheet erosion after brief storms in southwestern Iowa.

storms or melting snow. For example, in the Palouse region of western Idaho and eastern Washington where gullying is a problem, most erosion occurs in late winter and early spring: "Melting snow saturates the soil surface. Beneath the saturated layer, the ground is still frozen. Water is unable to permeate the iceblocked pores. Rain causes the already saturated soil to become a paste, creeping down the hillsides. If enough soil erosion occurs, the thick, deep scars of gullying appear."[5]

At one time, gullied land was common throughout the nation, particularly in the Southeast and the Midwest. Now, gullying continues to be a problem, but in fewer areas—west Tennessee and the Palouse hills of the Northwest for example.

Erosion is also caused by wind. When a field is fallow, the soil loses the binding effect that plant roots give it and becomes less

---

more awkward, but more accurate, sentence structure of "his or her." In addition, the use of the term *farmers* throughout the book may include ranchers.

THE NATURE AND EXTENT OF SOIL EROSION

Rill erosion.

cohesive. If rainfall or irrigation is inadequate, the soil becomes dry, light, and powdery. When winds blow across the field's surface, the detached soil particles become airborne. Heavier sand particles can even drop back to scour the earth of more soil as they bounce along.

The most famous episodes of wind erosion in the United States were the great "black blizzards" of the 1930s, which blew soil and dust hundreds of miles across the land. While wind erosion of that magnitude is rare today, windstorms continue to cause considerable soil losses, particularly in areas of low rainfall or in times of prolonged drought on fallow soils. Wind erosion, for example, can be extreme in areas such as Lincoln County, Colorado, where farmers have recently plowed what historically has been grassland. As one farmer described the situation, "Some of this marginal land being plowed won't even grow grass. But they are plowing anything you can get a tractor over." Because Lincoln County averages only 10 to 12 inches of rainfall per year, there

Gully erosion.

have been dust storms reminiscent of the dust bowl of the 1930s.

Factors Influencing Erosion

How severe erosion may be depends on a great many factors: the force with which wind or water strikes the earth; the duration of the windstorm or rainstorm or the quantity of snowmelt; characteristics of soil such as depth, texture, structure, organic matter percentage, total pore space, and size of individual pores; the length and steepness of a field's slope; the amount of plant cover; and the type of farming practices and cropping systems used.[6] Of all of these, one of the least controllable is climate.

Climate

If the United States had a climate of gentle rains and mists like that of Great Britain, it would have few water-caused erosion problems. But the United States does not have such a climate. Many areas of the country have rainstorms of considerable strength and intensity. On a field not protected with vegetation, a sudden

THE NATURE AND EXTENT OF SOIL EROSION

This 1930s photograph shows dust blown by the wind from an Iowa field.

Figure 2.1. Isoerodent map (originally developed from data obtained in rainfall records of eastern states and extended to the Pacific Coast by an estimation procedure)

SOURCE: Wischmeier and Smith, 1978.

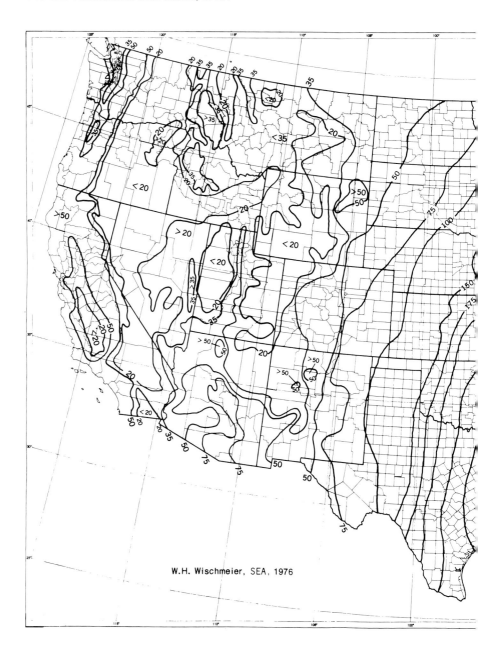

THE NATURE AND EXTENT OF SOIL EROSION

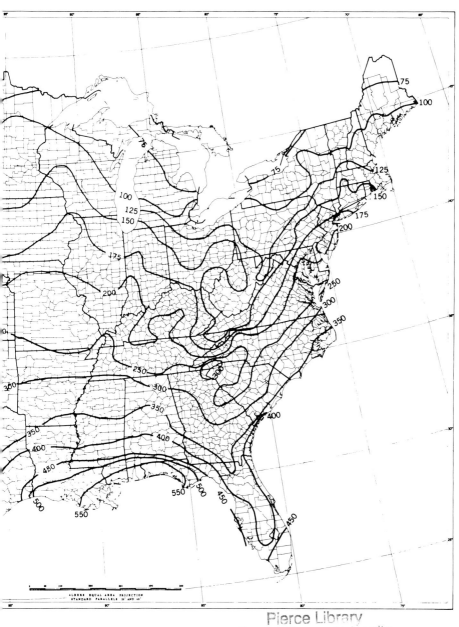

and severe rainstorm can erode many inches of soil in minutes.

The map in figure 2.1 shows the potential for soil erosion caused by rain. The lines identify areas with equal annual erosive potential from rainfall. Higher numbers mean greater potential. As the map indicates, there are great regional differences in erosion potential; the potential is far more severe in the southern states than in the northern states.

But potential erosion is not actual erosion. Actual erosion varies according to the type and amount of vegetative covering on the land during periods of heavy rains. In the Corn Belt region of Iowa and Illinois, for example, rains frequently occur in June, when the land has been plowed and planted but the plants have not yet matured enough to provide protection for the soil. In the South, rains occur more frequently in the winter months, when the land may be protected by a wintertime cover crop. Thus, rain-induced erosion may be worse in the Corn Belt than in the South, despite the heavier rainfall in the South. In arid areas of the West, serious water erosion may exist because there is not enough rainfall throughout the year to establish plants as ground cover for the infrequent rainstorms that do occur.

A map of wind erosion potential would look different from a map of rain erosion potential. Winds tend to be most severe in the Great Plains, especially Oklahoma, Kansas, northwest Texas, and Colorado.

Soil Type and Topography

Soil type and topography also influence the severity of erosion. Soils in the United States range from poorly drained to very dry, from sandy to clayish, from acid to alkaline, and from shallow to deep. They also vary in slope characteristics, porosity, organic content, temperature, and the capacity to supply nutrients.

All other things being equal, land is better for crops if it is nearly level, with just enough slope for good drainage. About 45 percent of the cropland used in the United States falls into this category. Only 10 percent of croplands exceed slopes of 12 percent. (see table 2.1). This is not surprising, since it is expensive, difficult, and even dangerous to harvest more steeply sloping areas.

Yet even gently sloping land erodes. In the central Corn Belt, erosion from slopes of 2 to 4 percent is estimated to be 2.6 times greater than that from slopes of 0 to 2 percent. Erosion from slopes of 4 to 7 percent is estimated to be 6.2 times that of nearly

Table 2.1. Slope of agricultural land by use (in percentages)

Slope	Cropland	Pastureland and native pasture	Rangeland
Level and nearly level (0 to 2 percent slopes)	45	28	11
Gently sloping (2 to 6 percent slopes)	25	18	4
Sloping (6 to 12 percent slopes)	20	22	14
Moderately steep (12 to 20 percent slopes)	7	14	14
Steep (20 to 45 percent slopes)	2	10	26
Very steep (>45 percent slopes)	1	8	31

SOURCE: USDA-RCA, 1980 appraisal, Pt. I, p. 2–28.

level land.[7]

Under a system devised by the Soil Conservation Service, U.S. soils are grouped into 8 land-capability classes. Soils in classes I–III are generally suited for frequent cultivation; soils in classes IV–VIII are severely limited in their usefulness for cultivation. Figure 2.2 summarizes these classifications.

Besides soil type, the classification system accounts for such physical attributes as natural drainage capacity and water supply. Capability subclasses are assigned by adding lower-case letters to the roman-numeral designation. For example, "e" for erosion potential, "s" for shallow or stony topsoil (that is, root zone limitation), "w" for moisture content and drainage problems, and "c" for climatic (cold and/or dry) limitations. Thus, class IIw lands are those of moderate-to-high fertility but have moisture content and drainage problems.

Why the states in the geographical center of the United States have a reputation for being excellent agricultural producers is apparent in figure 2.3, which depicts percentages of each soil classification by region. Well over half of the soil in the Northern Plains, the Corn Belt, and the Lake and Delta states is in the first 3 soil classes. In contrast, the southern states have considerable percentages of land in the higher-numbered (lower-quality) classes,

Figure 2.2. Soil Classification

				GRAZING			CULTIVATION			
LAND CAPABILITY CLASS	Wildlife	Forestry	Limited	Moderate	Intense	Limited	Moderate	Intense	Very Intense
I	▓	▓	▓	▓	▓	▓	▓	▓	▓
II	▓	▓	▓	▓	▓	▓	▓	▓	
III	▓	▓	▓	▓	▓	▓	▓		
IV	▓	▓	▓	▓	▓	▓			
V	▓	▓	▓	▓	▓				
VI	▓	▓	▓	▓					
VII	▓	▓	▓						
VIII	▓								

INCREASED INTENSITY OF LAND USE →

INCREASED LIMITATIONS AND HAZARDS ↓
DECREASED ADAPTABILITY AND FREEDOM OF CHOICE OF USES ↓

Shaded portion shows uses for which classes are suitable

Class I: Soils in this class have very few limitations on their use. They have the capacity for intense cultivation as well as grazing, rangeland, woodland and wildlife preserves. These lands are favorable for cultivation because, though they are generally high in native fertility, they are also responsive to fertilizer application, and have good drainage and water holding capacity.

Class II: These soils have some limitations that restrict their use. Though they may be of moderate to high fertility, they may not be able to sustain intensive cropping without some conservation practices. The following are some possible limiting factors: gently sloping lands, topsoil shallowness, imperfect drainage, or the presence of leached salts.

Class III: Soils in this class have severe limitations on their use. Crop cover must be maintained very carefully to protect land productivity. Limitations may include steep slopes, shallow soil root zone layers, high erosion hazards, excessive moisture content (slow permeability), low native fertility, alkali or saline conditions, or unstable soil structure. Still these lands have the capability for some cultivation as well as grazing, range-land, forestry and wildlife.

Class IV: This class is the last group of soils deemed suitable for cultivation. There may be severe restrictions on crop choice because of some of the following limitations: very steep slopes, severe erosion hazards, or shallow, saline, alkali, stony or waterlogged soils.

Class V: These soils are generally unsuitable for cultivation because of limitations other than erosion hazards. Their use may be impaired by frequent flooding, a short growing season, or excessively stony, damp or saline soils. These lands are generally suitable for pasture, rangeland, forestry or wildlife.

Class VI: These soils are also unsuitable for cultivation but erosion may be included as a factor limiting their use as well as the factors limiting Class V. This land is still suitable for pasture lands.

Class VII: These soils are unsuitable for cultivation or pasture lands. Their use should be restricted to range grazing, forestry or wildlife. The soils in this class are generally stony, on steep slopes, severely eroded, and of low native fertility.

(Figure from H.O. Buckman and N.C. Brady, *The Nature and Property of Soils*. New York, N.Y.: Macmillan, 1974. Text adapted from same.)

THE NATURE AND EXTENT OF SOIL EROSION

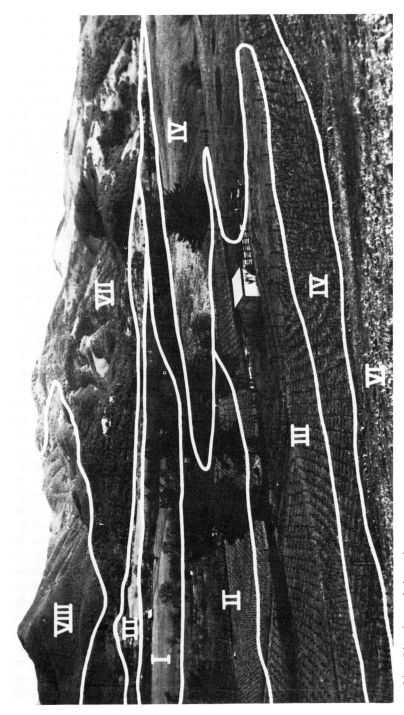

Profile of land-capability classes.

much of which is also designated as having erosion potential.

Farming Practices

A farmer's choice of cropping practices can also greatly influence the severity of erosion. On almost every field, the use of conventional plowing, in straight rows regardless of the topography* with all plant cover and crop residue removed, results in more erosion than if other techniques are used.

There are numerous management strategies and farm practices available to reduce erosion. First, and most obvious, the farmer can use his best land for crops and put any other land to different uses. After that he can introduce any number of conservation practices, such as strip cropping, contour planting, crop rotation, crop-residue retention on soil surfaces, terracing, waterway stabilization, and various conservation tillage practices. (See chapter 4 for a more detailed discussion of these techniques.)

Measuring Erosion Losses

How much soil is actually lost each year through erosion? That is a difficult question to answer.

Soil losses per acre are estimated by either the universal soil loss equation (USLE) or the wind erosion equation (WEE), both of which estimate the average annual tons of soil lost from each soil type as a function of climate, topography, cropping systems, and management practices. The equations have been developed from field experiments in various parts of the country. (A more detailed description appears in the appendix.)

There are a number of limitations in using the USLE and the WEE. Both measure the movement of soil but do not indicate whether the soil moves a few inches or a few miles. That is, soil losses measured by the USLE or the WEE may ultimately be deposited in a neighboring furrow, field, or river. Thus, the equations do not actually estimate soil loss *per se*; rather, they estimate the amount of dislodged soil. If USLE estimates are used as proxies for soil loss from a field, they may therefore overestimate the severity of erosion. This was the case in a study of an irregularly sloped watershed in Iowa, where it was estimated that 40

---

*This is also referred to as up-and-down-the-hill.

THE NATURE AND EXTENT OF SOIL EROSION

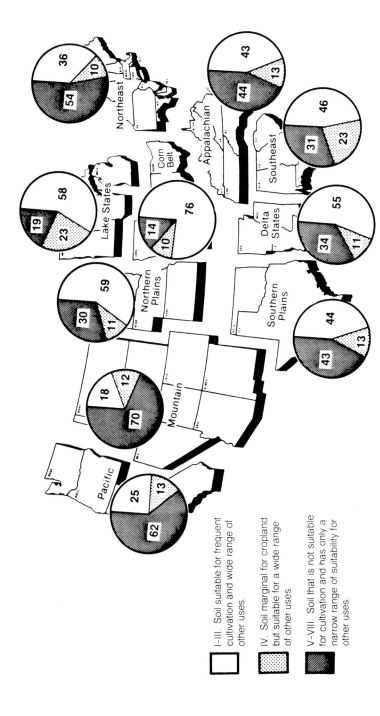

Figure 2.3. Distribution of soils by capability classes (percentages) and by farm production regions.

SOURCE: S. S. Batie and R. G. Healy, eds., *The Future of American Agriculture as a Strategic Resource* (Washington, D.C.: The Conservation Foundation, 1980).

percent of the soil displaced from the upper slope by rain runoff was deposited on the lower slope.[8] Overall, as much as 75 percent of transported soil may eventually be deposited on the same field from which it was dislodged.[9]

However, even where soil is redeposited on the same field, there appears to be some loss in productivity in the form of lost organic matter. That is, the lighter, nutrient-rich organic particles are more apt to be carried the farthest and thus are the most likely to leave the field. Also, some phosphorus, nitrogen, and potassium will be dissolved in water and carried away with the runoff. One study in Dodge County, Wisconsin, for example, found that soil formed by the deposit of material eroded from uphill sites was generally less productive than the soil it buried, because the nutrient-rich organic portions of the uphill soil had moved into waterways.[10]

Another limitation is that the USLE does not measure soil losses due to snowmelt, which can be substantial in some regions of the country. The Palouse, Nez Perce, and Columbia Plateau areas of the Northwest, for example, are not reflected in calculations of the USLE; yet losses in these areas can be as high as 50 to 100 tons per acre per year.[11]

Because of these limitations, USLE and WEE estimates need to be used with caution. Nevertheless, until more refined erosion estimates have been developed, these instruments are the best available for determining erosion rates.*

The soil losses measured by the USLE or WEE are average figures from measurements taken over an extended period of time and are usually reported in tons per acre per year. The computed losses are frequently related to soil-loss tolerances, or T-values, which are defined as the maximum annual soil losses that can be sustained without adversely affecting the productivity of the land. Soil erosion in amounts greater than T-values is frequently defined as "excess" soil erosion. For soils in the United States, the U.S. Department of Agriculture (USDA) has assigned T-values that usually range from 1 to 5 tons per acre, depending on the properties of the soil.

---

*Recently a modified version of the USLE, MUSLE, was developed. It attempts to overcome some of these limitations. MUSLE is used to measure the movement of the soil and not just the dislodging of soil as is the case with the USLE. See the appendix for more elaboration.

The validity of these numbers in representing maximum sustainable soil losses is doubtful, however,[12] A soil loss of 5 tons per acre per year translates into a net loss of 1 inch of soil every 30 years, or 1 foot of soil every 360 years.[13] On many soils, current T-values have been set too high to assure the long-term maintenance of the soil. Conversely, there is evidence that some soils may develop at relatively rapid rates. One study estimated that some Illinois soils would, with modern farming techniques, regenerate at not less than 12 tons per acre per year.[14]

But the relevant issue is not really that of soil reformation; rather, the important issue is protection of long-run productivity. If T-values are to represent maintenance of long-run productivity, they need to reflect the impact of technology on crop yields. If T-values also are to represent economic conditions, they have to reflect the costs and benefits of soil maintenance as well. For, as Crosson notes, what "if the minimum cost of achieving the T-values exceeds the cost of the productivity loss. . .?"[15] Because these future technological and economic influences are uncertain, however, their incorporation into T-values would be difficult and somewhat ambiguous. Such incorporation would probably mean lower T-values for numerous fields.

While cognizant of these arguments, many experts argue for retaining the concept of T-values as a physical measure of the maximum erosion allowable without reducing present soil depths. The reason is essentially an ethical one: that today's generation is only a temporary custodian of our soil resources.[16] R. Neil Sampson, in his book *Farmland or Wasteland: A Time to Choose*, is a good spokesman for this view.

> In short, there are serious questions about the concept of "tolerable" soil loss. Work is underway today to address those questions, but what is important to the public is the concept that should undergird that work. The objective of humans in using soils should be that those soils will maintain their productive potential despite that use. Any goal short of that is simply transferring the cost of today's excesses on to our children or grandchildren.[17]

There are also political problems with changing T-values. Larry Vance, chief of Ohio's Division of Soil and Water Districts and formerly director of Iowa's Department of Soil Conservation, summarized these dimensions of the issue: "If we changed T-values, a new farmer probably wouldn't know the difference. Some guy, who is supposed to know, will still be telling him 'here are some acres that will lose productivity over time.' But

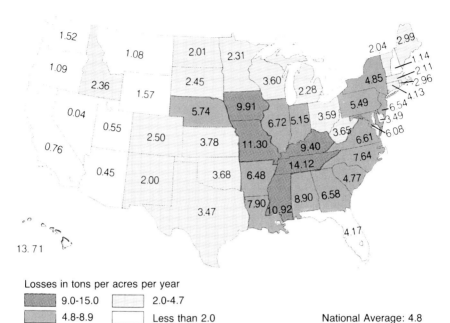

Figure 2.4. Sheet and rill erosion from water on cropland, by state, 1977.
SOURCE: *Environmental Trends* (Washington, D.C.: Council on Environmental Quality, 1981).

to an older farmer, changing T-values may suggest to him that he has invested money unwisely. We have something called the 4-ton club out in Iowa. They have taken care of their land for many years. Spent a lot of time and money. And they feel good about it. All of a sudden, they get the impression it really wasn't important after all. I'd say then you've got a real credibility problem on your hands."[18]

Erosion Losses

The national average loss of soil on croplands from water erosion, based on the USLE, was estimated at 4.8 tons per acre in 1977. Tennessee headed the list, with an estimated loss of 14.12 tons per acre, followed by Hawaii, with a loss of 13.71 tons per acre, Missouri, with 11.30 tons per acre, Mississippi, with 10.92 tons per acre, and Iowa, with 9.91 tons per acre (see figure 2.4). Wind erosion losses in 10 Great Plains states averaged 5.3 tons per acre in 1977. Texas led, with an average loss of 14.9 tons per acre, followed by New Mexico, with a loss of 11.5 tons per acre, and Colorado, with 8.9 tons per acre (see figure 2.5).

THE NATURE AND EXTENT OF SOIL EROSION

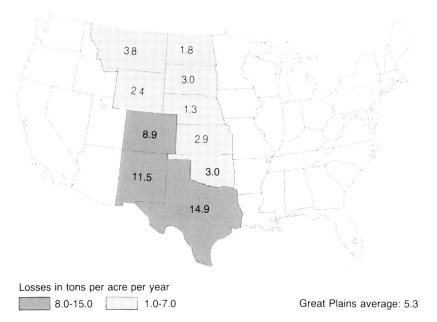

Figure 2.5. Wind erosion on cropland in the Great Plains states, 1977.

SOURCE: *Environmental Trends* (Washington, D.C.: Council on Environmental Quality, 1981).

If wind erosion losses are added to those of water erosion, Texas is the state with the greatest cropland losses. Tennessee, Hawaii, and New Mexico are close behind.

National average numbers do not give a full picture of erosion, for there are considerable regional differences. Most erosion occurs on specific and relatively small proportions of the total cropland acreage.

According to the 1980 RCA appraisal, 4 regions have severe water-caused erosion problems: the Palouse and Nez Perce prairies and the Columbia Plateau of west central Idaho, eastern Washington, and north central Oregon; the loess (or wind-deposited) soils of Nebraska, Kansas, Iowa, and Missouri; the silty uplands of the southern Mississippi Valley, which includes portions of Tennessee, Kentucky, Mississippi, Louisiana, and Arkansas; and the intensely cultivated areas of Aroostook County, Maine.[19]

The Palouse, Nez Perce, and the Columbia Plateau areas contain some of the world's best nonirrigated wheat lands. Barley, peas, and lentils are also grown there. Although precipitation is low, the steep slopes and lack of plant cover can lead to erosion rates of up to 100 tons per acre per year.[20]

Table 2.2—Combined sheet, rill, and wind erosion on land used for row crops and small grains*, 1977

Tons of soil eroded per acre per year	Acres x 1,000	Tons of excess erosion [1] x 1,000	Percent of acres	Percent of excess erosion
0 - 4.9	203,247	—	60.2	—
5 - 9.9	67,152	133,487	19.9	8.5
10 - 14.9	24,976	180,444	7.4	11.4
15 - 19.9	13,162	162,112	3.9	10.3
20 - 24.9	7,557	131,322	2.2	8.3
25 - 29.9	5,610	125,269	1.7	8.0
30 - 39.9	6,348	187,754	1.9	11.9
40 - 49.9	3,507	138,188	1.0	8.8
50 - 74.9	3,307	180,473	1.0	11.5
75 - 99.9	1,214	98,311	.4	6.2
100 - 149.9	643	73,428	.2	4.7
150 - 199.9	388	64,972	.1	4.1
200 & over	420	99,125	.1	6.3
TOTAL	337,531	1,574,885	100.0	100.0

*Includes summer fallow.

[1] Number of tons exceeding 5 tons per acre annually.

SOURCE: Computed from National Resource Inventory data, USDA, SCS, 1978.

The loess soils of Iowa, Nebraska, Missouri, and Kansas accumulated in prehistoric times. Although they may be as deep as 100 feet, they are extremely erodible, especially when planted in corn and soybeans, as they are now.

The southern Mississippi Valley is characterized by gently sloping bottomlands and steep valley sides. Much of this land is intensively cultivated for row crops such as corn and soybeans. Since the soil is highly erodible, this intensive use results in high soil losses.

The eastern portion of Aroostook County, Maine, is potato country. Slopes in this area used for crops may be as steep as 25 percent. It has been estimated that the upper 24 inches of soil have already been removed as a result of erosion.[21]

Even within regions, the most serious problems of soil erosion are concentrated on a relatively few acres. Table 2.2 shows the amount of combined sheet, rill, and wind erosion that exceeded 5 tons per acre per year on land cultivated in row crops and small grains in 1977. On over 60 percent of the cropland, combined annual rates of erosion did not exceed 5 tons per year. In contrast, almost 70 percent of the combined erosion over 5 tons per acre per year was concentrated on 8.6 percent of the cropland.

The USDA estimates that overall there are 1.2 billion acres with an annual soil loss of 5 tons per acre or less, 124 million acres with losses of 5 to 14 tons per acre per year, and 61 million acres with losses exceeding 14 tons per acre per year.

Are the Losses Increasing?

Whether soil erosion rates today represent an increase since the 1930s probably cannot be verified, despite citations to the contrary.* However, some research evidence suggests that soil erosion may be increasing. Studies conducted in Iowa reveal that, while annual soil losses decreased from 21.1 tons per acre in 1949 to 19.5 tons per acre in 1952 to 14.1 tons per acre in 1957, losses in 1974 were estimated to have increased to 17.2 tons per acre.[25]

What of the future? Are more losses expected? That will depend, in part, on how many acres are planted in crops. There are several

---

*For instance, the General Accounting Office report, *To Protect Tomorrow's Food Supply, Soil Conservation Needs Priority Attention*, quotes a 1972 Iowa State University report that the United States was losing 4 billion tons of soil a year through water erosion in 1972, compared with 3 billion tons in 1934, and that this translated into an average loss to farmers of 12 tons per acre annually in 1972, compared with 8 tons in 1934.[22]

The 4-billion-ton figure, which is widely quoted, appears to have as its original source a book entitled *Erosion and Sediment Pollution Control*, by R.P. Beasley.

> The loss of an estimated 4 billion tons of soil from land in the United States each year (more than ten times the amount of material excavated during the construction of the Panama Canal) affects many people, but primarily the landowner. It is estimated *that 3 billion tons of this total are lost from agricultural and forested lands.*[23]

The 3-billion-ton figure appears to come from the *1938 Yearbook of Agriculture: Soils and Man*.

> It has been estimated that some 3,000,000,000 tons of soil are washed annually from overgrazed pastures and cultivated or barren fields, to be poured into streams, harbors, reservoirs, lakes and oceans, or deposited on bottom lands and flood plains.[24]

As a careful reading shows, the erosion estimates for agricultural land in these sources are very similar. Thus, the validity of the GAO's comparison of 4 billion tons with 3 billion is doubtful, even if the estimates are assumed to be accurate.

studies predicting future cropland requirements. Generally, they are based on conservative estimates of improved yields and liberal estimates of the growth of food demands. According to these studies, estimates of additional cropland required by the year 2000 range from 26 million acres to 113 million acres.[26]

Whatever growth takes place, many of the additional acres put into crops would probably come from more erosive lands. If any of the projections prove to be true, then we can expect more soil erosion problems in the future unless the adoption of soil conservation strategies increases significantly.

References

1. H.O. Buckman and N. C. Brady, *The Nature and Properties of Soils* (New York, N.Y.: MacMillan Publishing Co., 1969), p. 299-303.
2. J.R. Williams, "The Influence of Soil Erosion on the Potential Productivity of Soil," U.S. Department of Agriculture, Science and Education Administration, viewpoint draft, presented to RCA Coordinating Committee July 18, 1980.
3. F.J. Stevenson, "Organic Matter Reactions Involving Herbicides in Soil," *Journal of Environmental Quality* 1 (1972): 333-43.
4. D.H. Simms, *The Soil Conservation Service* (New York: N.Y.: Praeger Publishers, 1970), p. 7.
5. S. Vira and H. Riehle, "Conservation in the Palouse: An Economic Dilemma," in Walter Jeske, ed., *Economics, Ethics and Ecology: Roots of Productive Conservation* (Ankeny, Iowa: Soil Conservation Society of America, 1981), p. 451.
6. R.N. Sampson, "Soil Degradation: Impacts on Agricultural Productivity," a report presented to the National Agricultural Lands Study by the National Association of Conservation Districts, 1980.
7. S.R. Aldrich, Assistant Director, Illinois Agricultural Experiment Station, testimony before the U.S. Congress, House Committee on Agriculture and House Committee on Science and Technology, *Joint Hearings on Agricultural Productivity and Environmental Quality*, 96th Cong., 1st sess., July 25, 26, 1979, p. 14-18.
8. R.F. Piest, R. G. Spomer, and P. R. Muhs, "A Profile of Soil Movement on a Cornfield," in *Soil Erosion: Prediction and Control*, Special Publication no. 21 (Ankeny, Iowa: Soil Conservation Society of America, 1977), p. 160-66.
9. R.N. Sampson, "Soil Degradation."
10. O.P. Engelstead, W.O. Shrader, and L.C. Dumenil, "The Effect of Soil Thickness on Corn Yield," *Proceedings of Soil Science Society of America* 25: (1961) 497-99.
11. U.S. Department of Agriculture, *Soil and Water Resources Conservation Act, 1980 Appraisal Part II, Soil, Water and Related Resources in the United States, Analysis of Resource Trends* (Review Draft) (Washington, D.C.: U.S.

Department of Agriculture, 1981), p. 37.

12. W.E. Larson, "Protecting the Soil Resource Base," *Journal of Soil and Water Conservation* 36 (1):13-16 (1981).

13. *Ibid.*, p. 15.

14. L. Bartelli, "Soil Development Deterioration and Regeneration," paper presented at the National Research Council, Soil Transformation and Productivity Workshop, Washington, D.C., October 16-17, 1980.

15. P. Crosson, *Productivity Effects of Cropland Erosion in the United States*. (Draft Manuscript) (Washington, D.C.: Resources For The Future, 1982), p. V-3.

16. Ibid, p. V-30.

17. R.N. Sampson, *Farmland or Wasteland: A Time to Choose* (Emmaus, Pa.: Rodale Press, 1981), p. 126.

18. Quoted in Ken Cook, "Commentary: Soil Loss–A Question of Values." *Journal of Soil and Water Conservation* 37(2): 92.

19. U.S. Department of Agriculture, *Soil and Water Resources Conservation Act, 1980 Appraisal, Part I, p. 107-08.*

20. *Ibid.*, p. 107-08.

21. *Ibid.*, p. 107.

22. U.S. Comptroller General, *To Protect Tomorrow's Food Supply, Soil Conservation Needs Priority Attention*, CED 77-30 (Washington, D.C.: General Accounting Office, 1977), p. 6.

23. R.P. Beasley, *Erosion and Sediment Pollution Control* (Ames, Iowa: Iowa State University Press, 1972), p. 14.

24. G. Hambridge, "Soils and Man—A Summary," in *1938 Yearbook of Agriculture: Soils and Man* (Washington, D.C.: U.S. Government Printing Office, 1938), p. 7.

25. J.F. Timmons, "Agriculture's Natural Resource Base: Demand and Supply Interactions, Problems, Remedies," in *Soil Conservation Policies: An Assessment* (Ankeny, Iowa: Soil Conservation Society of America, 1979), p. 55.

26. These studies include the following: Council on Environmental Quality, *National Agricultural Lands Study* (Washington, D.C.: U.S. Government Printing Office, 1981); M. Brewer and R. Boxley, "Agricultural Land: Adequacy of Acres, Concepts and Information," *Journal of Agricultural Economics* 63(5): 879-87 (1981); Council on Environmental Quality and the U.S. Department of State, *The Global 2000 Report to the President* (Washington, D.C.: U.S. Government Printing Office, 1980); appraisal data of the U.S. Department of Agriculture, Soil Conservation Service, *Natural Resources Inventory* (Washington, D.C.: U.S. Department of Agriculture, 1977); E.O. Heady, "The Adequacy of Agricultural Land: A Demand Supply Perspective", in P.R. Crosson, ed., *The Cropland Crisis: Myth or Reality?* (Baltimore, Md.: Johns Hopkins University Press, for Resources for the Future, 1982); and P. Crosson and S. Brubaker, *Resource and Environmental Effects of U.S. Agriculture* (Baltimore, Md.: Johns Hopkins University Press, for Resources for the Future, 1982).

Chapter 3

The Effects of Erosion

Soil erosion affects crop yields by stripping the land of its richest soil layers. That soil then adds dust to the air and sediment and chemical pollutants to lakes and rivers, impairing their quality and usefulness to society. It is no wonder there is growing concern over erosion. The effects of it can be destructive, long-lasting, and costly to reverse.

Yields and Erosion

It is difficult to generalize about the relationship between crop yields and soil erosion. Much of what is known is specific to given areas. While some broad conclusions can be drawn from specific studies, "average" measurements representing the entire nation are not very useful. For instance, the statement that if the average erosion loss in the Corn Belt is allowed to continue, potential yields of corn and soybeans will be reduced by 15 to 30 percent by the year 2030 masks many assumptions and many locational differences.[1] The danger of such broad generalizations is that they also mask the complexity of the problem and can lead to inappropriate recommendations.

Even the assessment of yield reductions as a result of soil erosion at a specific site is difficult, because any measurements must be taken while holding everything else constant. Over time, farmers may spread increasing amounts of fertilizer, both to substitute for eroded topsoil and to enhance the natural fertility of the soil. They may also change tillage practices, use different kinds of machinery, and grow improved varieties of plants. Since these changes generally result in larger yields, it is exceptionally difficult to isolate the effect of erosion.

Despite these complicating factors, scientists have approached the problem of measuring damage to soil productivity from erosion in several ways. One approach has been to find isolated field situations where a comparison can be made between eroded and noneroded farms under otherwise similar management systems and farm conditions. Another approach has been to simulate farm conditions in controlled experiments that measure changes in yields as a function of topsoil loss. Productivity indices are then developed that can be used to estimate yield reductions per acre-inch of topsoil loss for different soils and crops throughout the country. Other approaches include using linear programming or regression models to relate yields to soil depths.

The results of such studies have shown that a relationship exists between soil erosion and reduced yields on many soils. In some soils, farmers can improve their yields and compensate for the effects of erosion with fertilizers. This is most frequently the case on deep topsoils where erosion has depleted only soil nutrients. If erosion has reduced the water-holding capacity of the soil, the rooting depth, or the water infiltration rate, however, the addition of fertilizers may not offset the yield-reducing effects of erosion.[2] The crops most affected by erosion are wheat and the row crops of corn, cotton, and soybeans. Less affected are alfalfa, clover, and perennial grasses.

The ultimate effect of erosion on yields differs by soil type, by crop, and by management practices, as is illustrated by a case study of crop yields in west Tennessee.[3] This study showed that erosion adversely influenced yields, but did so quite differently in four different Tennessee soils: Memphis, Grenada, Loring, and Brandon.

Memphis, Loring, and Grenada are fine-silty soils formed of loess deposits greater than 4 feet in thickness. Memphis soils are well drained; Loring and Grenada soils are moderately drained and are characterized by fragipans in their lower subsoils that restrict root development as well as the movement of water and air. Brandon soils, many of which remain covered by forests, are fine-silty soils as well but are composed of relatively thin loess deposits (20 to 48 inches thick) over gravelly materials. The case study showed that soil erosion would have less effect on the Memphis soils than on any of the other 3 types. Severe erosion would reduce corn yields by 14 percent on Memphis soils, 22 percent on Grenada soils, 23 percent on Loring soils, and 30 percent on Brandon soils.*

Perhaps the most important distinction to consider relative to erosion's effects on yields is not the difference among soils *per se* but rather the cost of repairing or compensating for soil losses.

While actual cost estimates are not available, it is reasonable to assume that they will be higher when there are toxic materials in the soil or bedrock or fragipans beneath the remaining topsoil, all of which set limits on plant root depths and are difficult or impossible to overcome. This is estimated to be the case in 10 percent of the nation's croplands. Conversely, there are many situations where productivity losses attributable to erosion can be offset by increased use of fertilizers, improved plant varieties, or various farming techniques.

> [S]tatistical analysis of yield trends in major corn, soybean, and wheat growing counties from 1950 to 1980, indicates that the positive effects of technological change on productivity far outweighed the negative effects of erosion.[4]

Because much of erosion's effects in the past have been offset, and because of the use of new technologies, crop yields per acre have increased considerably in the last 30 years, as table 3.1 shows. Figure 3.1 puts those trends in historical perspective.

For over 200 years, agricultural productivity in the United States has grown almost continuously. This growth occurred slowly at first, with hand-power and animal-power technology. Then the technological and scientific inventions that followed World War I brought dramatic advances in productivity.

Crop yields increased in the past, not only because inputs were

---

*The estimated yields reported are realistic estimates of average yields attainable with current technology. The estimated yields were developed using the following assumptions:

1. Crops would be produced under natural rainfall conditions without irrigation.

2. Crops would be produced using all production practices recommended by the University of Tennessee Institute of Agriculture.

3. Crops would be produced in cropping systems that would hold soil losses within the tolerance limits set for the individual soils by the University of Tennessee Agricultural Experiment Station Bulletin 418, *Predicting Soil Losses in Tennessee under Different Management Systems*. Severely eroded soil was defined as soil that has a plow layer consisting largely of subsoil.

Table 3.1. Average yields of key crops in the United States

	Crop (bushels/acre)		
Year	Corn	Wheat	Soybeans
1930/39	24.2	13.3	16.1
1940/49	34.1	17.1	18.9
1970/79	89.6	31.4	28.1
1976/80	96.1	32.0	28.9

SOURCE: Adapted from P. Crosson, *Productivity Effects of Cropland Erosion in the United States* (Draft Manuscript) (Washington, D.C.: Resources for the Future, 1982), based on *Agricultural Statistics* (Washington, D.C.: U.S. Government Printing Office, 1972).

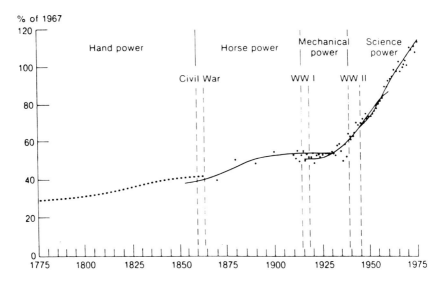

Figure 3.1. U.S. agricultural productivity growth during the past 200 years. SOURCE: K. R. Farrell, *Productivity in U.S. Agriculture*, ESS Report no. AGESS810422 (Washington, D.C.: U.S. Department of Agriculture, 1981).

more productive, but also because more inputs were employed. The main exception to this generalization is labor. The use of more powerful machinery, chemical fertilizers and pesticides, and irrigation has increased dramatically, while the number of hours spent in farm work has declined (see figure 3.2).

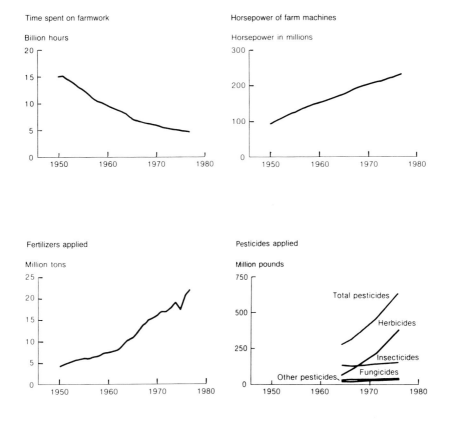

Figure 3.2. Agriculture inputs, 1950-1978.
SOURCE: *Environmental Trends* (Washington, D.C.: Council on Environmental Quality, 1981).

The reasons for these trends can be found in the relative prices of each type of input. From 1950 to 1978, gasoline prices rose twice as much as fertilizer prices, but labor prices rose more than twice as much as gasoline prices. Furthermore, land prices increased more than 4 times as much as gasoline prices.[5] The farmers have, therefore, substituted the relatively cheap inputs of fertilizer, water, and gasoline for the relatively expensive ones of land and labor. Thus, although soil has been eroding throughout the past decades, the increased production made possible by the expanding use of low-priced inputs has compensated, in terms of output, for the effects of that erosion.

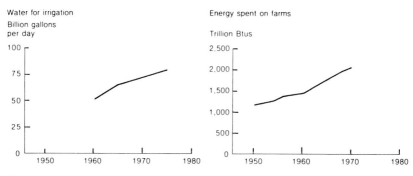

Figure 3.2. (cont.)

SOURCE: *Environmental Trends* (Washington, D.C.: Council on Environmental Quality, 1981).

Future Productivity

There is concern that such production increases will not continue in the near future. For one thing, inexpensive energy and plentiful water seem to be things of the past, and new avenues of inexpensive growth are not readily apparent.* For another, it appears that the growth of agricultural productivity has slowed. Table 3.2 shows that the average rate of change in total productivity has declined from 2.2 percent annually during the 1950-65 period to 1.8 percent annually during 1965-79.

One reason that some experts fear that cropland productivity growth will remain slow is because of insufficient funding for research.[6] They cite the large contribution that investment in research and education has made in the past to U.S. agricultural productivity growth and are disturbed about the lack of growth in government funding. If these researchers are right, greater demand for food and feed grains may be met only by greater use of inputs, including land.

Unless new technologies that can once again boost yields per acre are developed, technologically induced changes can no longer be counted on to substitute completely for the natural productivity of the land. The effects of soil erosion on yields will then

---

*Surface water rights on most rivers in the arid West, for example, are already over-appropriated, and the costs of groundwater withdrawal are climbing.

Table 3.2. Annual average rates of change (percent per year) in total outputs, inputs, and productivity in U.S. agriculture, 1870-1979.

Item	1870-1900	1900-1925	1925-1950	1950-1965	1965-1979
Farm output	2.9	0.9	1.6	1.7	2.1
Total inputs	1.9	1.1	0.2	−0.4	0.3
Total productivity	1.0	−0.2	1.3	2.2	1.8
Labor inputs	1.6	0.5	−1.7	−4.8	−3.8
Labor productivity	1.3	0.4	3.3	6.6	6.0
Land inputs	3.1	0.8	0.1	−0.9	0.9
Land productivity	−0.2	0.0	1.4	2.6	1.2

SOURCE: V. W. Ruttan, "Agricultural Research and the Future of American Agriculture," in S. Batie and R. Healy, eds., *The Future of American Agriculture as a Strategic Resource* (Washington, D.C.: The Conservation Foundation, 1980).

become more evident in those areas with little topsoil remaining.

Possible losses in productivity over the next decades could be rather high. One study recently completed by the U.S. Department of Agriculture (USDA), using yield data from 1,100 county surveys and the Iowa State University Linear Programming Model, estimated that, "if the current level of erosion were allowed to continue for the next 50 years on the 290 million acres contained in the USDA-RCA statistical model, erosion would cause a reduction in productive capacity equivalent to the loss of twenty-three million acres of cropland, or eight percent of the total base considered."[7] Such yield reductions were estimated to occur even if farmers intensified the use of fertilizers and used improved management practices.*

There are experts, however, who are optimistic about developing new technologies for the enhancement of agricultural productivity, including those no longer dependent on inexpensive water, energy, or land.[9] Many of these anticipated breakthroughs

---

*While this is probably the best estimate available on future productivity losses, it should be used with caution, since the data used in the model and the specification of the model were not what the researchers ideally would have liked to have had.[8]

will come, it is envisioned, from the biological sciences and plant and animal genetic research. If the predictions are accurate, a far lower future productivity "price to pay" for present erosion rates will result than if the more pessimistic visions of the future of American soil prove true.

It is clear that future production will be the result of numerous interacting factors. In addition to erosion rates, these include the costs and availability of soil substitutes, such as fertilizers and new technologies, the costs of soil rebuilding and the availability of such methods, and the management of soils, as well as the physical attributes of the plants and soils themselves.

Environmental Impacts

While farmers have reason to be concerned with the effect of erosion on their farms, there is no financial incentive for them to be sensitive to off-farm impacts. These effects—on air and water quality—are extremely difficult to quantify, but they are real nonetheless.

Erosion and Air Quality

Estimates of U.S. cropland's contribution to particulates in the air caused by wind erosion range from a low of 33 to a high of 239 million tons annually.[10] Even the lowest estimate is considerable. Emissions from point sources, such as smokestacks, contribute fewer than 20 million tons of particulates annually. Dust problems are particularly acute in the arid and semiarid areas of the Great Plains, far West, and Southwest, especially so during spring and fall plowing.

Erosion and Water Quality

Agriculture is thought to be the largest contributor to nonpoint water pollution.[11] Sediment from soil erosion and the water runoff carry along such pollutants as fertilizer residues, insecticides, herbicides and fungicides, dissolved minerals (such as salt), and animal waste-associated bacteria.*

Nutrients and Pesticides

Nutrients are chemical elements necessary for plant growth. They include nitrogen, phosphorus, and potassium. Nutrients are released naturally when organic matter decays; they may be leached from animal wastes; or they can be applied through chemical fertilizers. They may also cause pollution problems when they reach streams or other water bodies.

Large quantities of nutrients that end up in the nation's lakes and streams come from fertilizers. And most of the nation's croplands are heavily fertilized. Over 96 percent of the 1977-79 corn acreage received nitrogen fertilizers, 88 percent received phosphorus fertilizers, and 82 percent received potassium fertilizers. The comparable numbers for wheat were 63 percent, 42 percent, and 18 percent.[13] Because of this intensive use, farm fertilizer is thought to be the greatest source of both nitrogen and phosphorus in the nation's watersheds.[14]

Nutrients are carried off in sediment. In fact, because the fine soil particles carried away first have a higher capacity to absorb phosphorus and are carried farther in runoff, transported sediment is often richer in phosphorus and nitrogen than the original soil.[15] Nutrients can also dissolve and thus be present in water separate from sediment.

The amount of pesticides used on farmland has been increasing for the last several decades. In 1976, over 196 million acres of land were treated with herbicides, 75 million with insecticides, and 10.5 million with fungicides (some acres received more than one type of chemical).[16] Three crops—corn, soybeans, and cotton—accounted for 70 percent of all herbicide use on farms in 1979, and two crops—corn and cotton—accounted for nearly 64 percent of insecticide use. Fungicides are used primarily on fruit

---

*Cropland does *not* necessarily produce the greatest amount of erosion *per unit acre*. Sediment runoff from road construction in Virginia, for example, has reached 190 tons per acre. In Georgia, erosion rates from roadside cuts ranged from 42 to 289 tons per acre. Strip mining and urban development also contribute large sediment loads to streams. However, the large number of acres in cropland means that cropland produces more total sediment to streams and lakes than any other source.[12]

Fish kill.

and vegetable crops.[17]

Both nutrients and agricultural chemicals are generally washed away with sediments during storms, with a few severe storms producing most of the sediment yield. In one study, the 10 largest rainstorms produced a total soil loss of 7.8 tons per acre, while 62 smaller storms accounted for only 0.38 tons per acre.[18] Excess nutrients in water bodies can cause eutrophication, an enrichment of the waters. High levels of nutrients induce rapid growth of algae, which subsequently die and decay, reducing available oxygen in the water as well as releasing certain toxins.[19] The depletion of oxygen causes some plant and animal species to die out and otherwise disrupts biological activity. Excessive nutrients can also raise the costs of water purification for public water supplies, and the unpleasant sight and smell of rotting algal blooms can decrease a reservoir's recreational value.

High levels of agricultural chemicals in water can also be directly harmful to people, animals, fish, and plants. In 1975, for example, agricultural sources of pollution were considered responsible for 26 percent of the source-identified fish kills, second only to industrial causes. Of 118 agriculturally related fish kills, pesticides were thought to be responsible for 63.[20] The long-term effects of such chemicals as they build up in food chains are much harder to assess, although they have been definitely linked to reproductive failure in birds and fishes. (There are, however, agricultural chemicals that do not accumulate in the food chain and can be used under normal management practices with apparently little environmental damage.)

Not much is known yet about the impact of herbicides and fungicides on water bodies and the plants and animals that live in and around them. Studies have shown that some agricultural chemicals can be found in estuarine systems and even in the drinking supply of some counties.[21] Some of these chemicals have been thought responsible for creating harmful microbiological changes in soil, for causing reductions in the productivity of ecosystems, or for causing cancers.[22]

Dissolved Minerals

Runoff of minerals can also be a problem. In areas with extensive irrigation, salts can accumulate on the surface of the soil as a result of evaporation. In these areas, soil erosion can transport considerable amounts of salt, which, in large quantities, can be toxic to plants and fish. Salt-bearing water may also require substantial treatment before it is suitable for human consumption.

Sedimentation

Even if soil particles are free of excess nutrients, toxins, and other adhering substances, sediment flows may produce high levels of turbidity and the siltation of streams, lakes, and reservoirs.*

Turbidity, or cloudiness, in the water hinders the photosynthetic process that is the basis for all natural food production. It can also affect the respiration, growth, and spawning of fish and shellfish.

Siltation reduces the size of lakes and the depth of rivers and

---

*Streams do have natural sediment-carrying capacities. If water is artificially clarified, the stream will compensate by eroding its bed or banks to obtain more sediment.

This New Jersey coastal plain pond is surrounded by former potato land and suffers from heavy weed growth. Siltation caused some of the blockage, but excessive runoff containing fertilizer residues completed the job.

streams. A study of Winyah Bay, South Carolina, found that over 1 million tons of sediment per year were deposited in the bay, resulting in the need for substantial dredging to maintain navigational channels. Much of the sedimentation came from agricultural sources in the bay's watershed. Sedimentation in the Winyah Bay ship channel has become such a problem that the U.S. Army Corps of Engineers is considering relocating the port.[23] Siltation of lakes and rivers can lead to increased flooding problems and the deposition of infertile sediment on productive lands. It can also create a need for added storage capacity for dam reservoirs, since the volume of water that can be stored is reduced as sediment builds up behind a dam.

Extracting excess sediment from water for municipal uses in-

Barge full of silt and dredged material taken from river in an effort to improve navigation.

creases costs to water treatment facilities. Silt in water also increases costs to industrial users, putting wear on turbines and industrial cooling equipment and increasing cleaning and maintenance costs. Too much sediment, turbidity, and chemicals can ruin an area's suitability for water-skiing, sports fishing, swimming, and even camping.

Siltation of waterways is not without benefits, however. Many of the nation's wetlands, deltas, and beaches are the result of siltation. Silt built the extensive delta and wetlands that exist today at the mouth of the Mississippi River. In fact, there is concern that the Mississippi Delta is being starved because upstream dams and reservoirs are capturing sediment and what sediment the river does carry is often deposited too far past the wetlands to sustain delta formation.[24]

No one knows for sure how much it costs each year to undo

Aerial photograph of the Mississippi River in the vicinity of Head of Passes, showing extensive wetlands.

the damage done by siltation. One analysis made in 1980 estimated that sediment-related damages total nearly $500 million yearly.[25] Another estimated that, in 1976, $240 million was spent by the U.S. Army Corps of Engineers for dredging channels, harbors, reservoirs, ditches, streams, and lakes.[26] Such aggregated figures must be suspect, however, because accurate data per year by regions are virtually nonexistent.

The Case of Reelfoot Lake, Tennessee. Reelfoot Lake in Tennessee is a dramatic case of the substantial problems that can result from agricultural runoff. Located in the northwestern part of the state in Lake and Obion counties, Reelfoot was spectacularly formed by a large earthquake in 1811:

> The night of December 15, 1811 . . . was clear and quiet and there was nothing unusual that would give the least sign or omen of the coming catastrophe. About two o'clock in the morning of December 16 the inhabitants were suddenly awakened by an awful noise. . . . Landslides swept down the steeper bluffs and hillsides, larger areas of land were uplifted and still larger areas were sunk. . . . The land where Reelfoot Lake is now situated was sunk from a depth of one to fifty feet. . . . After, it was said that the Mississippi River, in filling up these sunken lands, ran upstream for a period of forty-eight hours.[27]

"[T]hese sunken lands" became an irregularly shaped, 50,000-acre lake. It is relatively shallow, with a present maximum depth of only 25 feet. Most of the lake is only 8 to 9 feet deep. The lake drains over 163,600 acres, of which 62,258 are now in row crops, part of which lie west of the lake in a flat, rich plain stretching to the Mississippi. The area east of the lake is characterized by rolling hills.

For many years, farming was not profitable in the eastern portion of the watershed. Now, soybeans are grown in the erosive soils there, and sedimentation of the lake has become a major problem. Some newly constructed recreational areas have been closed because boat docks that were useful only a year ago have been blocked by sediments. One dock, in use 2 years ago, is now 20 yards inland from the shoreline.

With the sediment has come chemical pollution. Rangers at Reelfoot are concerned that these chemicals will harm the bald eagles, golden eagles, and ospreys that depend on the lake's fish for food. Some claim the bird populations have already dwindled because of the pesticides. In addition, there have also been suspected pesticide-caused fish kills in Reelfoot Lake. The lake is used not only for sport fishing but occasionally for commercial fishing as well.

Excursion boats continue to provide scenic "nature tours" of Reelfoot, but the surface of the lake is increasingly difficult to negotiate because of shallows and weeds. What was once one of the nation's largest inland crane rookeries is now almost dry.

Reelfoot is literally disappearing under the increasing rates of agriculturally related sedimentation. It now has an area of only 18,000 acres, and it will be filled entirely by the year 2032 if present rates of sedimentation continue.[28]

References

1. U.S. Department of Agriculture, *Soil and Water Resources Conservation Act, 1980 Appraisal, Part II, Soil, Water, and Related Resources in the United States: Status, Conditions and Trends* (Review Draft) (Washington, D.C.: U.S. Department of Agriculture, 1980), p. 3-13.

2. W.D. Shrader and G.W. Langdale, "Effect of Soil Erosion on Soil Productivity." Iowa Agricultural and Home Economics Experimental Station Journal Paper no. J-9600 (Ames, Iowa: Iowa State University).

3. G.J. Buntley and F.F. Bell, *Yield Estimates for the Major Crops Grown on the Soils of West Tennessee*, Bulletin 561, (Knoxville, Tenn.: University of Tennessee Agricultural Experiment Station, 1976).

4. P.R. Crosson, *Productivity Effects of Cropland Erosion in the United States*, (Draft Manuscript) (Washington, D.C.: Resources for the Future, 1982), p. IV-30.

5. O.C. Doering III, "Energy Dependence and the Future of American Agriculture," in S.S. Batie and R.G. Healy, eds., *The Future of American Agriculture as a Strategic Resource* (Washington, D.C.: The Conservation Foundation, 1980), p. 191-224.

6. For further discussion of productivity trends see E.O. Heady, "The Adequacy of Agricultural Land: A Demand-Supply Perspective" in P. Crosson, ed., *The Cropland Crisis: Myth or Reality?* (Baltimore, Md.: Johns Hopkins University Press, for Resources for the Future, 1982); V.W. Ruttan, "Agricultural Research and the Future of American Agriculture," in Batie and Healy, eds., *The Future of American Agriculture as a Strategic Resource*, p. 117-56; and P.R. Crosson and S. Brubaker, *Resource and Environmental Effects of U.S. Agriculture* (Baltimore, Md: Johns Hopkins University Press, for Resources for the Future, 1982).

7. National Association of Conservation Districts, *Soil Degradation: Effects on Agricultural Productivity*. National Agricultural Lands Study Interim Report, no. 4. (Washington, D.C.: National Association of Conservation Districts, 1980), p. 26-27.

8. Crosson, *Productivity Effects of Cropland Erosion in the United States*, p. IV-11–15.

9. S. Wittwer, "New Technology, Agricultural Productivity, and Conservation," in H.G. Halcrow, E.O. Heady, and M.L. Cotner, eds., *Soil Conservation*

Policies, Institutions and Incentives (Ankeny, Iowa: Soil Conservation Society of America, 1982), p. 201-15.

10. U.S. Environmental Protection Agency, Interagency Energy/Environment R&D Program, *Setting Priorities for Control of Fugitive Particulate Emissions from Open Sources*, EPA 600/7-79-186 (Washington, D.C.: U.S. Environmental Protection Agency, August 1979).

11. U.S. Department of Agriculture, Soil Conservation Service, *America's Soil and Water Condition and Trends* (Washington, D.C.: U.S. Department of Agriculture, 1980), p. 30.

12. W.H. Wischmeier, "Cropland Erosion and Sedimentation," in B.A. Stewart, D.A. Woolhiser, W.H. Wischmeier, J.H. Caro, and M.H. Frere, *Control of Water Pollution from Cropland, Vol. II: An Overview*, ARS-H-5-2 (Washington, D.C.: U.S. Department of Agriculture, 1976).

13. Crosson and Brubaker, *Resource and Environmental Effects of U.S. Agriculture*.

14. B.A. Stewart, D.A. Woolhiser, W.H. Wischmeier, J.H. Caro, and M.H. Trere, *Control of Water Pollution from Cropland, Vol. I: A Manual for Guidelines Development*, ARS-H-5-1 (Washington, D.C.: U.S. Department of Agriculture, 1975).

15. M.H. Frere, D.A. Woolhiser, J.H. Caro, B.A. Stewart, and W.H. Wischmeier, "Control of Non-Point Water Pollution from Agriculture: Some Concepts," *Journal of Soil and Water Conservation* 32(6): 263-4 (1977).

16. Crosson and Brubaker, *Resource and Environmental Effects of U.S. Agriculture*.

17. D. Pimentel and S. Pimentel, "Ecological Aspects of Agricultural Policy," *Natural Resources Journal* 20(3):555-586 (1980).

18. D.A. Haith and R.C. Loehr, eds., *Effectiveness of Soil and Water Conservation Practices for Pollution Control*, EPA-600/3-79-106 (Athens, Ga.: U.S. Environmental Protection Agency, Environmental Research Laboratory, Office of Research and Development, 1979).

19. M H. Frere, "Nutrient Aspects of Pollution from Cropland," in B.A. Stewart, D.A. Woolhiser, W.H. Wischmeier, J.H. Caro, and M.H. Frere, *Control of Water Pollution from Cropland, Vol. II: An Overview*, ARS-H-5-2 (Washington, D.C.: U.S. Department of Agriculture, 1976).

20. U.S. Environmental Protection Agency, *Fish Kills Caused by Pollution in 1975*, EPA-44/9-77-004 (Washington, D.C.: U.S. Environmental Protection Agency, 1977).

21. National Research Council, Committee on Agriculture and the Environment, *Productive Agriculture and a Quality Environment* (Washington, D.C.: National Academy of Sciences, 1974).

22. P. Crosson, *Conservation Tillage and Conventional Tillage: A Comparative Assessment* (Ankeny, Iowa: Soil Conservation Society of America, 1981) p. 31.

23. J. Clark, *Winyah Bay Reconnaisance Study* (Summary Report) (Washington, D.C.: The Conservation Foundation, 1980), p. 25, 49.

24. S.M. Gagliano and J.L. van Beek, "Mississippi River Sediment as a Resource," in R.S. Saxena, ed., *Modern Mississippi Delta Depositional Environments and Processes*. This guidebook was prepared for the AAPG/SEPM

field trip held in conjunction with annual AAPG/SEPM convention, New Orleans, Louisiana, May 23-26, 1976, and has been reprinted by Coastal Environments Inc., 1260 Main Street, Baton Rouge, LA 70802.

25. Pimental and Pimental, "Ecological Aspects of Agricultural Policy."

26. The Council for Agricultural Science and Technology (CAST), *Socioeconomic Implications*, Report no. 92 (Ames, Iowa: CAST 250 Memorial Union, 50011, January 1982) p. 9.

27. State of Tennessee, Department of Parks and Recreation, *Tennessee Reelfoot Lake Map and Guide* (Nashville, Tenn.: Department of Parks and Recreation).

28. Gassner, Nathan, Brown Architects/Planners, Inc., *Master Plan for Reelfoot Lake State Park*, prepared for the state of Tennessee, Department of Conservation, November 1969.

Chapter 4

Techniques for Reducing Soil Erosion

Farmers have numerous management techniques available to them to reduce erosion. Not all techniques are suitable on all lands, but in most cases the choice of the right techniques can substantially reduce soil loss.

The most obvious choice is that of the location of the farm itself. A farm located on the rich, well-drained soils of Hanford fine, sandy loam in California is preferable to a farm located in the steeply sloping Honeoye silt loams of New York. Most farmers, however, are committed to their locations; for them, moving is not a viable option for soil conservation. Instead, they must look to some of the many other techniques available to reduce soil losses. These include changing the characteristics of the soil and topography, choosing appropriate crops, rotating crops, constructing terraces or waterways, and using conservation tillage methods.

Changing the Characteristics of Soil and Topography

If the soil and topography of a farm are not what the farmer would like, there are limited possibilities for change. If the soil has suitable water-holding capacity, soil fertility can be enhanced through the addition of fertilizers. Wet soils can be drained by the addition of underground tiles or ditching. Lands can be leveled in a process referred to as land forming. Steep lands can be terraced.

Choosing Where to Plant

Few farms are composed of fields of identical soil classifications. A farmer can elect to plant crops on the least erosive lands, or, as the Soil Conservation Service's motto states, "Plant the best, save the rest."

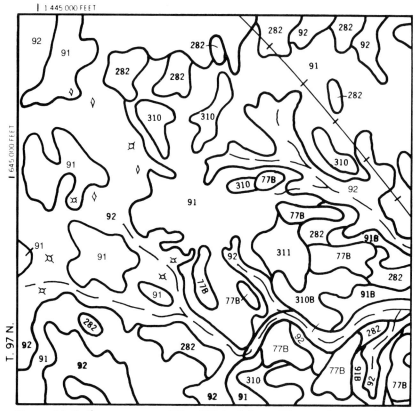

Figure 4.1. Soil survey map of Jim O'Brien's farm. The numbers on the map are indicative of different soil types and classes.
SOURCE: USDA, Soil Conservation Service.

Figure 4.1 is a soil-survey map of an Iowa farm. One conservation strategy available to the farmer is to plant crops continuously only in class I and II soils, leaving the lower quality soils as pasture or woods. The farmer may need to correct the limitations of a field, such as by installing tiles. By draining the wet field to make it highly productive, the farmer may be more willing and financially able to leave the class III and IV lands in grass.

Crop Rotation

Crop rotation involves alternating soil-conserving crops with soil-depleting crops. Row crops, such as corn and soybeans, are often soil depleting. Many legumes and sod-forming grasses are soil conserving—that is, they improve soil structure and reduce erosion. Thus, when rotations are made between the two types of

TECHNIQUES FOR REDUCING SOIL EROSION

Table 4.1. Cropping Systems and Soil Erosion

Crop	Average annual loss of soil per acre (tons)*	Percentage of total rainfall running off land
Bare, cultivated, no crop	41.0	30
Continuous corn	19.7	29
Continuous wheat	10.1	23
Rotation of corn, wheat, and clover	2.7	14
Continuous bluegrass	0.3	12

*The average of 14 years of measurement of runoff and erosion at Missouri Experiment Station, Columbia.

SOURCE: M. F. Miller, *Cropping Systems in Relation to Erosion Control* (Columbia, Mo.: Missouri Agricultural Experiment Station Bulletin no. 366, 1936).

crops, the soil is improved during alternate years. Table 4.1 shows the difference in soil loss with different cropping systems for a case study in Missouri. It is clear that, in this case, grass or legume cover reduced erosion substantially. The erosion rate dropped from 19.7 tons annually per acre with continuous corn cultivation to 2.7 tons per acre with a rotation of corn, wheat, and clover.

Strip Cropping

Strip cropping involves growing crops in a systematic arrangement of strips or bands that can serve as buffers. The strips can follow the contours of the land and may involve, for instance, a combination of corn and alfalfa hay. Alfalfa strips catch soil runoff from the strips planted in corn and serve as minor windbreaks as well. Because the velocity of water over vegetation is approximately one-sixth that of water over bare soil, strip vegetation slows erosion and encourages water filtration. A 100- to 125-foot strip across a moderate slope will filter about 75 percent of the sediment from normal field runoff.[1]

Retaining Crop Residues

Soil can also be conserved by allowing the residue from crops to remain on the soil surface. The presence of residue may increase the infiltration rate. One researcher estimates that for every ton of residue per acre, soil loss from water is reduced by

Corn and alfalfa strips planted on the contour in Iowa, 1956.

Minimum-till planting in corn stubble on a farm in Nebraska, 1979.

65 percent.[2] Other researchers find that if two tons of residue per acre are left on the land, erosion is effectively controlled.[3]

Reducing Wind Erosion

Special soil conservation practices combat wind erosion. They include reducing the width of the field by barriers or strip crops at right angles to the prevailing wind, irrigating, using windbreaks of trees and shrubs, or simply keeping crop cover or residue on the fields. These techniques can reduce particulates from escaping into the air by as little as 5 percent for windbreaks to almost 100 percent for cover crops.[4]

Construction Measures

Locating crops on the best soils, rotating crops, strip cropping, retaining crop residues, and using the techniques for reducing wind erosion can be relatively low-cost methods of achieving soil conservation goals. This is particularly true in those cases where the farmer does not have to plant much land in low-value crops to achieve the desired level of conservation.

Construction measures, in contrast, are generally more expensive, and some require considerable expertise. Still, these methods can be very effective in reducing soil losses.

For example, terraces can reduce soil loss on some landscapes. Terracing breaks the length of a slope into shorter segments, which reduce water velocity and slow water runoff in the terrace channels. However, terraces must be carefully designed to accommodate large modern farm machinery.

Creating waterways and stabilizing drainage paths are also effective techniques. There are several ways these systems can be used to control soil loss. One is constructing waterways covered with vegetation. Depending on the volume of water and sediment to be controlled, the waterways may be of various widths and contain various types and densities of vegetation. Large terrace-like waterways, often covered with grass, can channel water away from a problem area. These diversions are often constructed above gullies and sloping fields. Buried pipes can also be used to channel excess water from terraces. Subsurface drains lead water to an outlet where it can be disposed of effectively.

Sediment basins are another technique. They are created by barriers placed across waterways to catch silt. When water velocity slows at the barrier, sediment drops out.

Fencing pastures and carefully siting livestock watering troughs to control the concentration of animals per acre can reduce overgrazing, which has been a frequent cause of soil erosion in the West.

Conservation Tillage Practices

Conservation tillage practices are cultivation techniques that reduce the exposure of soil to the potentially erosive effects of wind and water. They frequently differ considerably from more conventional tillage techniques.

Conventional tillage has traditionally meant early plowing (even as early as the preceding fall) to prepare the field for planting. Such a field is exposed to storms and wind throughout the winter months. Generally, the field is plowed again before planting. The spring tillage may be "up and down" the fields, with little regard to field slopes. Not until plants have grown high enough to provide a "canopy" is the soil protected from erosion.

Conventional tillage generally also means removing most plant

TECHNIQUES FOR REDUCING SOIL EROSION 61

This air view, taken the day after a 4-inch rain, shows water being held above terraces. There were 36 miles of terraces on this farm, and no terrace failure resulted from the storm.

Waterway on a farm in Washington.

residue from the surface of the field at harvest. While perhaps aesthetically pleasing and neat in appearance, a clean field has a high erosion potential.

Conservation tillage practices range from modifications of traditional farming techniques to practices that depend on the late 20th-century technology.

Perhaps the simplest conservation "tillage" practice is to eliminate fall plowing altogether. A farmer who waits until spring to plow can often substantially reduce annual erosion losses.

Contour plowing is one of the oldest techniques of conservation tillage. It simply means plowing around, rather than up and down, a hill. When plowing and planting are done on the contours of the hills, the furrows provide means of catching soil runoff and may increase water infiltration. Contour cropping can reduce

TECHNIQUES FOR REDUCING SOIL EROSION 63

Sediment basin, at the base of the hill, used for erosion control. Note the contour plowing and strip-till techniques.

Contour rows reducing water runoff in a Texas field.

erosion by as much as one-half over up-and-down-the-hill plowing.[5]

No-till and *minimum-till* are practices that have been getting considerable attention in recent years. These labels cover a range of techniques that forego conventional plowing and field preparation in favor of disturbing the soil as little as possible.

No-till eliminates almost all tillage. There is no seedbed preparation before planting. Usually, residue cover is retained and cut through only to plant seeds. Chemicals—not conventional plowing—are used to control weeds.

Ridge-plant involves the development of ridges with residue left in the furrows between them. Seeds are planted in the ridges. The residue in the furrows collects the runoff, and eroded sediment is thus deposited right next to its source.

Plow-plant eliminates secondary tillage. Seeds are planted in

TECHNIQUES FOR REDUCING SOIL EROSION 65

This field in Alabama was cultivated up-and-down-the-hill; terraces were not maintained, nor were outlets kept open. Most of this damage occurred from rain on May 18, 1969.

a plowed field with no secondary tillage or seedbed preparation. This technique reduces the time that soil is exposed before plants begin developing.

Till-plant involves opening a seed furrow, dropping in the seed, and closing the soil over the seed, all in one process. It is an effective erosion control system when rows are laid out along the contours of a field.

Chisel plowing loosens the soil for air and water flow without inverting it. A specially designed chisel plow cuts the soil with pointed shanks, which are then pulled through the soil. Some plant residue remains on the surface. This technique minimizes the exposure of subsurface soils yet provides for enhanced root growth. It is especially useful where the soil has become very compact from the use of large farm machinery.

No-till equipment preparing to plant soybeans in wheat stubble.

Strip-tillage uses rotary tillers spaced to provide a broad seedbed. Seeds are dropped into the furrows and covered immediately.

Sweep-tillage lifts and breaks the soil to kill weeds but leaves residues in place to enhance water infiltration.

Listing pushes soil into ridges, drops seeds into the furrows between, and covers them—all in one operation. When used on the contour, the ridges catch water and slow runoff.

Wheel-track planting involves plowing to invert the soil, then dropping seeds into the tractor wheel tracks. This method eliminates secondary tillage or seedbed preparation.

Several studies have examined the relative effectiveness of these tillage alternatives. In one study, no-till with residue cover reduced erosion 63 percent over conventional tillage involving moldboard plowing, secondary preparation, and no residues. Conventional tillage with residue left in the field after harvest resulted in a 14 percent reduction in erosion.[6]

Another study compared chisel-plow, till-plant, and no-till sys-

TECHNIQUES FOR REDUCING SOIL EROSION

Soybeans planted (no-till) in wheat straw.

tems with conventional tilling under simulated rainfall conditions of Indiana and Illinois. Chisel plowing would have decreased soil loss by 94 percent; had till-plant been used, the loss reduction would have been 60 percent; had no-till been used, the loss reduction was calculated at 85 percent (all were compared with conventional tillage after a high-intensity storm).[7]

In many cases, conservation tillage offers the farmer advantages beyond reduced soil loss. Less tillage, less soil exposure, and more residue usually cause the soil to retain more moisture. That means reduced demand for irrigation waters in many areas. Also, less tillage gives a farmer more flexibility in land use, since row crops can be cultivated on steeper slopes with less danger of soil loss. Also, fuel requirements are usually reduced, because fewer trips need to be made across the field. Fertilizer requirements can be reduced, because less tillage means plant nutrients

A one-person plow-plant rig.

are left in place. In the long run, soil quality actually may be improved by the addition of more organic matter, the retention of more moisture, and the reduction of soil compaction.

However, conservation tillage techniques usually require the use of increasing amounts of herbicides and pesticides to control weeds and insects.* No-till methods sometimes increase herbicide use by a factor of 2.5.[8] As a Tennessee district conservationist stated, "A farmer has to be a good 'chemical' man to be successful with no-till." The farmer using conservation tillage practices faces some new problems. He must consider the increased costs of chemicals. In addition, residual herbicide buildup may damage future crops. For example, the herbicides used with corn can kill soybeans. Corn-soybean rotation may thus be constrained if

---

*Some farmers claim that they find little difference in their use of chemicals to control weeds and insects with conservation tillage as opposed to conventional tillage. This may imply that they were over-applying the chemicals when using conventional tillage techniques.

An example of conservation tillage. This photograph shows an Iowa farmer cultivating his corn crop without removing soybean residue.

excessive amounts of herbicides remain in the soil between plantings. Furthermore, increased herbicide use is not always successful in controlling weeds. And increased chemical use may result in increased chemical contamination of water.

The farmer has another kind of problem, too. A no-till or minimum-till farm may look "peculiar." Rather than neat fields of rich, dark earth lined with straight rows of crops, the fields are littered with decaying stalks and leaves and spotted with weeds. One Illinois farmer described the conflict a farmer feels:

> A lot of the conservation tillage that is started in the fall is defeated in the spring. In the fall of the year, instead of moldboard plowing, some of these farmers will come in with a chisel plow—and it looks trashy, you know, it just looks terrible. So when spring comes they'll come and hit it with a disk and just pulverize that trash, where it doesn't look like they've done any conservation tillage whatsoever.[9]

Conservation Tillage Trends

A district conservationist in west Tennessee observed that the drought of 1980 helped him convince farmers that there were

A winged-shovel chisel plow hooked in tandem with a stalk chopper.

positive returns to no-till in the west Tennessee area. The no-till cover maintained the scarce moisture longer than the conventionally tilled fields. and the result was increased yields. As one Tennessee farmer observed, "We couldn't see the difference between no-till and conventional tillage as much before 'cause of the rain. But now with this drought, yields are greater on those fields which have been no-tilled. I got 20 bushels an acre more with no-till." He went on, "Furthermore, we can see the problems of eroded soil more clearly. I have a field where most of the topsoil is gone on one end. I got 5 bushels an acre of corn there this year—what with the drought—but I got 100 bushels of corn on the other end of the field."

Observations like these are encouraging many U.S. farmers to adopt conservation tillage. It is now used on approximately 25 percent of the 387 million acres of land in crops. While this adoption rate has largely been in response to economic incentives that reduce labor and fuel requirements, the result is also less erosion than if conventional tillage practices had been followed.

Pierre Crosson of Resources for the Future predicts that, by the year 2010, 50 to 60 percent of the nation's cropland will be farmed with conservation tillage. (He also estimates, however, that the amount of cropland may increase by as much as 75 million acres and that much of that additional land would be more erosive than land now in crop production.[10] Thus, high levels of erosion will probably still occur.) Crosson believes that the lack of effective weed control is the main factor limiting adoption of conservation tillage.

Still, conservation tillage is clearly gaining attention, as the following "un-till" pledge promoted by a conservation committee in Jones County, Iowa, suggests: "I will till no soil UNTIL its time. If I don't have a darn good reason for fall tillage, I will wait UNTIL spring. If it's planting time and I still don't have a good reason, I will try NO-TILL."

References

1. B.A. Stewart, D. A. Woolhiser, W. H. Wischmeier, J. H. Caro, and M. H. Frere, *Control of Water Pollution from Cropland, Volume II: An Overview,* Report no. ARS-H-5-2 (Washington, D.C.: U.S. Department of Agriculture, 1976).

2. R. B. Held and M. Clawson, *Soil Conservation in Perspective* (Baltimore, Md.: Johns Hopkins University Press, for Resources for the Future, 1965).

3. J. V. Mannering, J. D. Meyer, and L. D. Meyer, "The Effects of Various Rates of Surface Mulch on Infiltration and Erosion," *Proceedings of Soil Science Society of America* (1963) 27:84-86.

4. U.S. Environmental Protection Agency, *Guidelines for Air Quality Maintenance Planning and Analysis, Vol. 3: Control Strategies,* EPA-450/4-74-003 (Research Triangle Park, N.C.: U.S. Environmental Protection Agency, 1974).

5. Held and Clawson, *Soil Conservation in Perspective.*

6. W. E. Larson, "Crop Residues: Energy Production or Erosion Control?" *Journal of Soil and Water Conservation* (1979) 34(2):74-76.

7. D. C. Griffith, J. V. Mannering, and W. C. Moldenhauer, "Conservation Tillage in the Eastern Corn Belt," *Journal of Soil and Water Conservation* (1977) 32(1):20-28.

8. O. L. Bennett, "Conservation Tillage in the Northeast," *Journal of Soil and Water Conservation* (1977) 32(1):9-12.

9. M. Lenehan, "Will the Corn Fields End Up In the Rivers," *The Atlantic,* December 1981, p. 25.

10. P. Crosson, *Conservation Tillage and Conventional Tillage: A Comparative Assessment* (Ankeny, Iowa: Soil Conservation Society of America: 1981), p. 31.

Chapter 5

Factors Affecting Farmers' Adoption of Conservation Practices

On a recent visit to the United States, a delegation from the People's Republic of China was confused when it was unable to obtain clear and simple answers to such questions as the following: Who picks the seed the farmer will use? Who decides which youths will become farmers? Who decides which farming practices are best?

The questions are amusing to Americans, who understand that farmers themselves ultimately make these decisions and receive most of the consequent rewards or burdens. This is true of conservation practices, too, and many factors influence a farmer's decision about what practices (if any) to adopt.

Some farmers do not see that they have soil erosion problems, even if government officials or neighbors suggest otherwise. Others recognize the problems but do not want to make the recommended changes—because they cost too much or because the farmer does not believe the changes will help, dislikes government's assistance, or is simply unwilling to try new techniques.

Other farmers may adopt conservation practices whether or not they are sound financial investments. A farmer with a strong land-use ethic, for example, may adopt conservation practices because it is the "right thing to do," regardless of whether he will personally reap the benefits of the investment. Some farmers follow totally the recommendations of Soil Conservation Service (SCS) officials, as does one farmer in west Tennessee: "I just tell the district conservationist to do what needs to be done and leave it to him." Nevertheless, since only 40 percent of America's cropland has been judged by the SCS to be adequately protected from soil erosion, clearly there are barriers to installing soil conservation practices. Some barriers are raised by the high costs involved, the investment incentives in rising land prices, land tenure arrangements, and tax and loan policies. Some simply stem from a farmer's preferences and beliefs.

Personal Preferences

The Judeo-Christian heritage, which stresses human dominion over the earth but has no commandment for husbanding resources, may fuel the psychology of those farmers who lack commitment to soil conservation techniques.[1] This purported lack of land ethic, combined with a frontier psychology that exalts independence, hard work, and absolute private property rights, could account for some of the failure to conserve. Furthermore, many farmers just do not perceive the same need for soil erosion control as do soil conservation specialists.

Several researchers have found that failure to see the need for erosion control practices was a significant factor in explaining the failure to adopt them. In one survey in Nebraska, "the SCS classified 82 percent of the farms as having a major soil erosion problem while only 2 percent of the operators and none of the landlords classified their farms similarly. Moreover, 54 percent of the operators and 55 percent of the landlords indicated either no or few erosion problems; yet SCS classified only 4 percent as having no problem."[2]

The authors of the study suggest that the different perceptions come partially from different perspectives. "SCS classifies soil erosion problems in terms of the amount of soil movement. Operators are more likely to classify problems in terms of the difficulties caused by soil erosion, the visibility of soil movement and the short-run effect of erosion on the economic, physical and operational aspects of farming."[3] In general, a greater proportion of younger operators agreed with SCS advice than did older operators.

Other factors also inhibit conservation practices. Some farmers want clean fields and straight rows no matter what advice they are given. Others have little confidence in the effectiveness of the soil conservation practices being recommended and therefore do not adopt them. Still others dislike government involvement and requirements. As one Tennessee farmer stated, "What erosion practices are needed, I do myself. I don't much like all that government red tape and specifications."

When farmers do adopt conservation practices, they choose those most compatible with their lifestyles. For example, some recommended conservation techniques require adding livestock to a farm. But livestock, which requires daily attention, would cramp the style of some Iowa farmers, who joke about the current crop rotation as being corn, soybeans, and Florida. That is, some farmers prefer a winter vacation in Florida after crop harvesting to tending hogs and cattle. This personal preference is reinforced when livestock prices are low in relation to those of crops. Also, farmers have differing backgrounds, educations, experiences, managerial skills, planning horizons, and attitudes toward risk taking. All of these factors substantially influence a farmer's willingness to adopt soil conservation practices.[4]

Cost of Conservation

Another reason conservation practices are not adopted is that many do not pay for themselves. The business-minded farmer, who must remain competitive to remain in farming, is simply not interested in such practices if they are not profitable. Even a farmer with a strong land ethic and a desire to practice soil conservation may find it is financially impossible to do so.

Practicing conservation involves changes, all of which have costs. No-till cropping, for example, may require the purchase of new equipment and will probably require more agricultural chemicals. Elimination of fall plowing means that a farmer must spend more time in the spring, when time is very valuable, preparing for planting. Adding forage crops as part of a conservation plan usually means adding livestock; doing so requires fences, specialized equipment, and local markets for feed, antibiotics, veterinary care, and other inputs.

Also, as any farmer can testify, farmers buy and own farms—not fields. If part of a farm includes inferior, more erodible soil classes surrounded by better soils, the farmer may find it uneconomical, as well as inconvenient, to treat the more erosive lands differently. If the farm is growing corn and soybeans, for example, the farmer will not normally find it practical to deal with an "island" of erosive soil by fencing it and using it for livestock, or sowing different crops on it, or letting it lie idle. Ownership boundary lines can also interfere with contouring or terracing. It is not surprising that farm layout is a significant obstacle to voluntary erosion control.[5]

One project in Tennessee illustrates the types of expenses that can be incurred in soil conservation efforts. In September 1979, the Gibson County, Tennessee, Rural Development Committee, in cooperation with the state Rural Development Committee, sponsored Operation SOS—Save Our Soil. Gibson County has one of the highest soil erosion rates in the nation, with averages of as much as 40 tons per acre per year and extremes as high as 100 tons per acre. It is said that the soil in Gibson County acts like sugar in the rain. Even relatively flat land can be deeply scarred with gullies after a severe rainfall.

Operation SOS was a demonstration project to promote methods of controlling soil erosion in west Tennessee. The project used volunteer funds and labor in an impressive effort to introduce conservation practices on 5 farms in only 1 day. Charles Harrison's farm, for example, received 5,180 feet of terraces, 3 sediment basins, 2 grassed waterways, and grading and stabilization structures. The total cost for the 37-acre project was $17,885, or $482 per acre. Harrison's farm, according to some of the planners, looks "better than any of the other SOS farms" and has had its annual erosion rates reduced from 26 tons per acre to 2.8 tons per acre. But the costs are significant in a county where farmland sells for $1,200 to $1,300 an acre. One farmer said he would not adopt anything like the "maximum" plan used on the Harrison farm. "Why, you would have more in working the land than the land cost originally."

For the business-minded farmer, a decision to maintain, improve, or deplete soil is an investment decision. Because of the competitive nature of farm wholesale markets, farmers have little control over the prices they receive for their harvests. Thus, from the farmer's point of view, profits depend on the size of the harvest and the costs of production. If a farmer is "mining" the soil—that is, if erosion rates exceed the rate at which soil is being formed—the ultimate result will probably be lower yields. Farmers can improve yields, however, through fertilizing, liming, irrigating, and other management techniques. The choice between a conservation system and a depletive system is thus basically a choice of whether to incur costs now or in the future.

Reducing erosion—through the use of contour plowing or terracing, or a change in crop rotation pattern—generally raises production costs and may entail a sacrifice in current income. Allowing erosion to continue, on the other hand, may eventually lower soil productivity and crop yields—and raise future production costs.

Thus, in a conservation system of farming, production costs will be higher initially than those of a depletive system, reflecting the investment in soil conservation. The production costs of a depletive system will probably rise over time (assuming no other changes) and can eventually exceed the cost of a conservation system, as the loss in revenues from lower yields and higher fertilizer costs becomes greater than the increased cost attributable to soil-conserving practices.*

For some tracts of land, conservation may be economical for the farmer when the land is first cultivated. This might be true on land that is fairly flat, where the topsoil is shallow but highly productive, and the subsoil is of substantially poorer quality. In this situation, losing an inch of topsoil could reduce yields dramatically; erosion could be reduced inexpensively if, for example, contour plowing or residue retention were used.

On other tracts of land, where conservation requires major land-moving technologies to form terraces and where straight-row, highly erosive cropping patterns can bring a high dollar return, private economics may dictate mining the soil. This is particularly true where the original topsoil is very deep. In western Tennessee, for example, some farms have as many as 80 feet of topsoil. Lands with topsoil of this depth could be eroded at the rate of 40 tons per acre for approximately 3,600 years before normal plowing would mix subsoil with topsoil. As one district conservationist lamented, "How do you talk soil conservation to a farmer who is sitting on 80 feet of topsoil?"

Because soil is literally an income-earning asset, the farmer's conservation decision becomes one of balancing possible losses of future income if topsoil is not maintained with the costs of maintaining the topsoil today. The question for the farmer is not whether to mine the resource, but how quickly. Other factors also influence this decision. For example, the lower the price of soil substitutes, such as fertilizers, the less likely the farmer is to conserve. Also, the lower a farmer's current net income, the less likely he is to conserve, since substituting future income for present income is financially impossible.

Several researchers have studied the return on conservation investments. Most suggest that many conservation practices do

---

*In some cases, the result of having followed a long-term depletive system may be so damaging to soil productivity that restoration to an earlier state is either impractical or impossible.

not yield a profitable return to farmers. One study examined 6 watersheds in western Illinois.[6] For a 20-year planning horizon, incomes were calculated under combinations of 3 conservation practices (contour plowing, straight-up-and-down plowing, and terracing), 3 tillage practices (conventional, plow plant, and chisel-plow), and rotations that ranged from continuous corn to a wheat-meadow mixture. These hypothetical operations resulted in soil losses ranging from 0.4 tons per acre annually to over 60 tons per acre. Average yearly income was calculated for each combination. These calculations took into account yield reductions from erosion but did not allow for the application of extra fertilizer.*

In general, yield reductions were very slight over the 20-year period, and incomes were higher with the use of more erosive rotations and tillage techniques. For example, in the Hambaugh-Martin Watershed, rotations away from continuous corn reduced income. That is, the extra income from continuous corn more than offset any gains from saving the soil by using crop rotations.

Significantly, however, the highest returns did not come from the most depletive practices. The maximum income per acre was achieved with continuous corn, up-and-down planting, and chisel plowing. Soil losses were reduced from 60.53 tons per acre, with continuous corn, up-and-down planting, and conventional plowing, to 9.42 tons per acre when chisel plowing was substituted. Further reductions in soil loss per acre to 7.64 tons were achieved for only pennies per acre when the farmer switched to contour plowing. In some instances, contouring and terracing resulted in the highest income, but even in these cases the difference in income between conservation and straight-row planting was very slight. Never did a shift in rotation from continuous corn to a less-profitable crop mix increase income. Yet the mixed-rotation pattern consistently resulted in a dramatic drop in soil loss. Nevertheless, the study generally showed the financial incentives for conservation (assuming no public assistance) either to be so small as to be insignificant or to be negative.

Another study, focusing only on the costs and benefits to farms

---

*Dismissing adjustments in fertilizer use makes erosion damages clearer but biases calculations for income downward if replacing the lost productivity with fertilizers would be cheaper than reducing the erosion.

using terraces on Illinois soils, supported the contention that some practices are not profitable: "Considering only the direct benefits from terracing, most farmers would lose personal income by investing in terraces."[7] This was the case despite the government's paying 50 percent of the construction costs of the terraces.

Another study examined the costs and benefits of soil conservation in the Southern Iowa Conservancy District.[8] Costs of erosion were expressed as yield reductions, as increased fertilizer inputs to maintain productivity at levels recommended by soil tests (but which never brought yields back to preeroded levels), and as increased fuel requirements associated with the poorer tilth of eroded soil. Information was available on changes in yields, fuel requirements, and fertilizers for slightly, moderately, or severely eroded soils. The cost of meeting erosion-tolerance limits was estimated by choosing the least costly method of achieving the limit from among 6 alternative management practices, all of which included a change in rotation that lowered the acreage of row crops and, therefore, resulted in a substantial loss of income. One alternative included terracing, which involved a major capital expenditure and raised production costs significantly. However, it was evident that meeting erosion-rate limits in this study area could not be achieved only by the use of the less costly methods of conservation tillage, residue management, and contouring. The authors concluded that, given the alternative conservation practices they considered, the cost of reducing soil erosion (assuming no public assistance) to "tolerable levels" was "three times as expensive as the benefits."[9]

None of these studies considered no-till practices. Although no-till involves the purchase of new machinery, in some cases it does pay. For example, one researcher found that, on well-drained soil in Ohio, yields increased as much as 10 percent using no-till.[10] On poorly drained soils, however, there were yield decreases. Another study found that no-till increased corn yields in Virginia as much as 36 percent.[11] However, two other researchers found that, at best, no-till would only equal conventional tillage yields in the climates of New York and New England.[12] While there do appear to be positive financial inducements for some farmers to try no-till practices, the payoff differs by soil type and amount of yearly rainfall.

Most studies of the costs of soil conservation consider a farmer's returns with and without soil-conserving practices. Usually, the soil-conserving practices are assumed to reduce soil losses to a single minimum level, such as 5 tons per acre. One recent study, however, analyzed, for 300 different soil classes, the difference in expected income for achieving each of 14 different erosion rates, ranging from 20 tons per acre per year to 0 tons.[13] The study showed that a farmer's net income changed little as he reduced soil erosion losses to 2 tons per acre. As he went from 2 to 0 tons per acre, however, net income was reduced dramatically.

When public costs of installing soil conservation practices are included, the same general relationships apply. A U.S. Department of Agriculture study estimates that the average total cost of reducing erosion to 10 tons per acre per year on those lands eroding at more than 14 tons per acre per year was less than $1 per ton. In contrast, the cost per ton of reducing erosion from 10 to 0 tons per year varies from $2.16 to $45.40 per ton.[14]

Clearly, the evidence is persuasive that as increased reductions in soil erosion are obtained, costs tend to increase. Thus, in highly eroding areas, considerable soil savings can be obtained at very little cost. However, reductions beyond a certain annual rate, generally from 2 to 9 tons per acre, come at substantially increased costs.

Also, the majority of studies conclude that most soil conservation practices are not economical investments for farmers. The major exceptions appear to be conservation tillage, contour plowing, and leaving residue. In some areas of the nation, these practices are effective in reducing erosion and do not decrease the profitability of farming; in some, they may even increase the profitability.

Land as an Investment

Decisions about soil conservation are complicated by the multiple reasons that exist for owning farmland. Production of food and fiber is only one. Investment is another. Investment in cropland has been an excellent inflation hedge. One study found that over half of the benefit from owning farmland in the years from 1920 to 1978 resulted not from what the land produced but rather from the contribution of the land to increases in the owner's net worth.[15] But the result of using land as an inflation hedge is often increased soil erosion.

Since the mid-1970s we have had a tremendous increase in the price of land and the cost of farm equipment. So we now have farmers who look upon farming as a real estate game. They buy the land and farm the hell out of it to meet their heavy payments, not worrying about preserving it because they believe it will keep going up in price ... It's like buying a rundown building in New York City and letting it deteriorate still further, knowing you'll still have the capital value in the end.[16]

This expectation of not being penalized heavily for erosion when land is resold appears to be reasonably accurate. Although there is some evidence that present *productivity* is a factor in current farm prices, there is little evidence that present *erosion rates* influence land prices, since erosion affects only future productivity. Agricultural Stabilization and Conservation Service (ASCS) and SCS officials in Texas, Tennessee, Iowa, and New Mexico when interviewed concurred that almost no premium is paid for land on which permanent conservation practices (such as terraces) have been introduced or on which the soil has been carefully husbanded.

If this is true, it may be because farmers lack knowledge on future yields, they do not believe that erosion will affect future crop yields, they predict future demand for crops will not lead to substantially increased prices, or they believe land-saving technologies will be available to compensate for any erosion-caused reductions in yields.[17] In addition, the perceived value of farmland as an investment may be so high that it dwarfs any considerations of yield differences when the land is used for crops.

For a realistic example, in 1978 (a reasonably "good" year for farmers) in Douglas County, Illinois, a farm could easily have had a selling price of $3,000 per cropland acre, only a part of which price could be accounted for by the land's expected revenues from farming. The remainder of the selling price represents the value of the land as investment. If a farmer had anticipated (1) an eventual yield of 20 more bushels per acre on his farm because it had careful soil stewardship, (2) an expected profit per bushel of 60 cents, and (3) the ability to invest money at 12 percent interest, then, all other things being equal, he should have been willing to pay a maximum of only $100 per acre more for the uneroded farm versus the eroded farm—not a large amount when compared with a selling price of $3,000 per acre.

This implicit calculation depends on the farmer's perception of the magnitude of difference in future yields on the husbanded versus the unprotected farm and on the expected future profits per bushel: the higher the perceived future-yield difference and profit per bushel, the more the farmer would be willing to pay for the husbanded farm relative to one unprotected. If the farm has only a few fields, or areas within fields, where conservation practices will actually result in yield improvements, then the farmer would be unwilling to pay significantly more for the protected farm. Similarly, if the conservation practices are on a farm that does not have potentially high erosion rates, whether or not the soil is protected, then the farm buyer will also be unwilling to pay much more for the protected farm.

An illustration of the difficulties of implementing soil conservation when land prices do not reflect future productivity levels was provided by a district conservationist in New Mexico. He was talking of a farmer who needed to reseed overgrazed lands:

> It'll take him 5 to 20 years to recover this cost, so you're doing something more for aesthetics than for the rancher's benefit. You're asking him to fix something that often times he didn't cause. Around here, you don't have the same land ownership year after year. If you bought the land—well, you know, there are folks who buy a ranch and strip it in 5 to 10 years and sell it. They make a large amount of money because the price of land goes up every year. They take all the best grazing off of it, and you come in and buy it, and somebody says, hey, your land's erosive and you're going to have to fix it. You're not going to like it, especially if the other guy's got the money and could afford it, and he's gone.

The evidence is mostly anecdotal as to whether or not farmers will respond to erosion differences on farms that are presently equally productive; there appears to be little or no research on why farmers respond as they do. The question, however, is crucial to the formulation of conservation policies.

Tenure Arrangements

Insecure property rights also work against soil conservation efforts. If a farmer cannot capture the future gains that arise from conservation decisions, he will have no incentive to conserve. Thus, if a farmer expects either to sell the farm shortly or to have his lease cancelled, he has little reason to begin conserving.

Encroaching urbanization can also lessen any incentives for conservation. Consider the case of the farmer growing turf grass near Richmond, Virginia. The grass is cut into long strips, rolled, and sold to nurseries to become instant lawns for residential homes. Each cutting removes a quarter inch of topsoil. When asked if he worried about depleting the soil, the farmer replied that although he figured that his farm would become unproductive by the year 2000, he would be ready to sell it as apartment building lots for the growing Richmond suburbs.

Leasing arrangements also may lessen conservation incentives.[18] Former Secretary of Agriculture Bob Berglund claimed that, "Our biggest problem is persuading absentee landlords to plow some of their earnings back into the soil."[19] A farmer in west Tennessee concurred:

> Absentee landlords are a bad problem. I lease 50 percent of my land, nearly 600 acres, and, like most of the land around here, it is either owned by widows of deceased farmers or some of the doctor bunch. I have put in terraces at my own expense on some of this land because the owners won't and because I think it is the right thing to do. But I'll tell you, some of my neighbors have criticized me for it.

In other circumstances, however, it is the landlord that blames the tenant for not incorporating soil conservation practices, despite encouragement. Nevertheless, it is an exceptional lease that financially encourages a tenant farmer to practice conservation on leased or rented land.*

Tax Policies

Tax policies can also encourage or discourage soil-depleting practices, even though such impacts are often unintended.

First, and maybe even foremost, in influence are the tax advantages that encourage farmland speculation: deductibility of interest on borrowed funds as a business expense, investment tax credit, several methods of computing accelerated depreciation,

---

*Another tenure-related concern has been the increase in corporate ownership of U.S. farmland. Some fear that a larger, more corporate agricultural system will lack a conservation ethic, and, because of their presumed drive for more return on investment, corporate owners will mine the soil more than noncorporate owners. The evidence is inconclusive. However, one study found no correlation between erosion rates and corporate or noncorporate ownership on a national level.[20]

and the treatment of any capital gains.

> The existence of a preferential capital gains tax in this market draws capital into real estate, not by the promise of higher earnings but by the promise of greater value retention. This distorts investment patterns, displaces operators whose focus must be on income flow rather than on net worth, and encourages patterns of land use that will minimize supervisory cost while waiting for land values to rise.[21].

Other tax policies encourage adoption of soil conservation practices by farmers in the higher tax brackets. Farmers have been allowed to consider certain investments in soil conservation practices as current expenses to be deducted in a single tax year, up to 25 percent of the gross income of the farm. Cost-sharing payments for conservation practices need not be included as individual gross income for federal tax purposes if they do not substantially increase annual farm income. However, the farmer may have to pay additional taxes if the fair market value of the land increases as a result of conservation improvement.[22]

Further, the 1976 Tax Reform Act and the Economic Recovery Act of 1981 reduce estate taxes and thus encourage intergenerational transfers. These tax laws would appear to encourage soil conservation if, as some observers claim, a farmer is more apt to practice conservation when he anticipates that his children will inherit and operate the farm.

However, some of these same tax policies provide financial incentives for the cultivation of land previously not cropped. For example, farmers can deduct costs for eradication of brush or construction of drainage and irrigation ditches.

Overall, however, the tax codes and regulations are so complex that in many cases they "could make it difficult for a landowner to evaluate the potential tax consequences of a conservation project. It seems likely that some people might avoid participating in a conservation project because of uncertainty about, or misunderstanding of, the tax laws."[23]

Property taxes appear to have little direct correlation with conservation practices.[24] However, they can put a significant dent in a farmer's pocketbook. A farm worth over half-a-million dollars might have a property tax obligation of, say, $5,000 a year in some states. An obligation of this size, although it can be deducted from federal taxes, could contribute to a cash-flow problem that causes the land-rich and money-poor farmer to need immediate returns on his farming investments. As several studies have found, liquidity and debt problems are significant constraints to adopting soil-conserving practices.[25] Clearly, in times of depressed crop prices, any financial obligations reduce the amount of money a farmer has available for conservation efforts.

Loan Policies

Few lending institutions appear to require any conservation practices on farmland as a contingent requirement for obtaining loan funds, unless such loans are specifically for funding conservation practices. Furthermore, the Farmers Home Administration offers loans to qualifying farmers below the rates of other financial institutions. In the past, the result of these phenomena is that land purchasers have literally been paid to borrow. This is the case, for example, if inflation is 12 percent per year, but the mortgage is only 9 percent per year. Such financing obviously makes farmland purchases very attractive.

In addition, the relationship of low interest rates on farm mortgages relative to the past high inflation means that agricultural land as an agricultural input has been historically undervalued and may have been treated accordingly. "This helps explain larger machinery allowing more land to be utilized per worker, larger acreage per farm, increased acreage farmed by owners and part-owners, a decline in the acreage farmed by full tenants, and a much lower increase in land productivity than in labor productivity during the 1970's."[26] If land is relatively inexpensive, there is less incentive to conserve soil than if the real price of land were higher.

References

1. H. Briemyer, *Farm Policy: 13 Essays* (Ames, Iowa: Iowa State University Press, 1977), p. 113.
2. H. Hoover and M. Wiitala, *Operator and Landlord Participation in the Maple Creek Watershed in Northeast Nebraska*, staff report NRED 80-4 (Washington, D.C.: U.S. Department of Agriculture, Economics, Statistics, and Cooperative Service, 1980), p. iii.
3. *Ibid.*, p. iv.
4. P. J. Nowak and P. F. Korsching, "Sociological Factors in the Adoption of Soil Conservation Practices," unpublished paper, Department of Sociology and Anthropology (Ames, Iowa: Iowa State University, 1981).
5. M. G. Blase and J. R. Timmons, "Soil Erosion Control in Western Iowa: Progress and Problems," Iowa State University Resarch Bulletin 498 (Ames, Iowa: Iowa State University, 1961).
6. A. S. Narayanan, M. T. Lee, K. Gunterman, W. D. Seitz, and E. R.

Swanson, *Economic Analysis of Erosion and Sedimentation; Mendota West Fork Watershed*, Department of Agricultural Economics, and Agricultural Experiment Station, AERR 126 (Urbana, Ill.: University of Illinois at Urbana-Champaign, 1974); M. T. Lee, A. S. Narayanan, K. Gunterman, and E. R. Swanson, *Economic Analysis of Erosion and Sedimentation; Hambaugh-Martin Watershed*, Department of Agricultural Economics, and Agricultural Experiment Station, AERR 127 (Urbana, Ill.: University of Illinois at Urbana-Champaign, 1974); A. E. Narayanan, M. T. Lee, and E. R. Swanson, *Economic Analysis of Erosion and Sedimentation; Crab Orchard Lake Watershed*, Department of Agricultural Economics, and Agricultural Experiment Station, AERR 128 (Urbana, Ill.: University of Illinois at Urbana-Champaign, 1974); M. T. Lee, A. S. Narayanan, and E. R. Swanson, *Economic Analysis of Erosion and Sedimentation; Seven Mile Creek Southwest Branch Watershed*, Department of Agricultural Economics, and Agricultural Experiment Station, AERR 130 (Urbana, Ill.: University of Illinois at Urbana-Champaign, 1974); A. S. Narayanan, M. T. Lee, and E. R. Swanson, *Economic Analysis of Erosion and Sedimentation; Lake Glenside Watershed*, Department of Agricultural Economics, and Agricultural Experiment Station, AERR 131 (Urbana, Ill.: University of Illinois at Urbana-Champaign, 1975); M. T. Lee, A. S. Narayanan, and E. R. Swanson, *Economic Analysis of Erosion and Sedimentation; Upper Embarras River Basin*, Department of Agricultural Economics, and Agricultural Experiment Station, AERR 135 (Urbana, Ill.: University of Illinois at Urbana-Champaign, 1975).

7. J. K, Mitchell, J. C. Brach, and E. R. Swanson, "Costs and Benefits of Terraces for Erosion Control," *Journal of Soil and Water Conservation* 35(5):233-236 (1980).

8. P. Rosenberry, R. Knutson, and L. Harman, "Predicting the Effects of Soil Depletion From Erosion," *Journal of Soil and Water Conservation* 35(3):131-134 (1980).

9. *Ibid.*, p. 134.

10. D. L. Forster, N. Rask, S. W. Bone, and B. W. Schurle, *Reduced Tillage Systems for Conservation and Profitability*, Department of Agricultural Economics and Rural Sociology, ESS 532 (Columbus, Ohio: Ohio State University, 1976).

11. C. Underwood, "No-till is the Word," *Water Research in Action* 1(3):1-3 (1976).

12. F. N. Swader, *No-plow Corn in New York*, proceedings of the Northeastern No-tillage Conference, Albany, New York, 1970. See also O. L. Bennett, "Conservation Tillage in the Northeast," *Journal of Soil and Water Conservation* 32(1):9-12 (1977).

13. J. F. Timmons and O. M. Amos, "Economics of Soil Erosion Control with Application to T-Values." Iowa Agricultural and Home Economics Experiment Station Journal Paper no. J-9625 (Ames, Iowa: Iowa State University, 1979).

14. U.S. Department of Agriculture, Agricultural Stabilization and Conservation Service, *National Summary Evaluation of the Agricultural Conservation Program: Phase I* (Washington, D.C.: U.S. Department of Agricul-

ture, 1981).

15. E. N. Castle and I. Hoch, "Farm Real Estate Price Components 1920-78," *American Journal of Agricultural Economics* 64(1):8-18 (1982).

16. Quoted in A. Crittenden, "Soil Erosion Threatens U.S. Farm Output," *New York Times*, October 26, 1980.

17. P. R. Crosson, "Diverging Interests in Soil Conservation and Water Quality: Society vs. the Farmer," paper presented at the annual meeting of the American Agricultural Economics Association, Clemson University, Clemson South Carolina, 1981.

18. The research available on tenure and leasing arrangements is not conclusive. See, for example, D. E. Ervin, "Soil Erosion Control on Owner-Operated and Rented Cropland," *Journal of Soil and Water Conservation* 37(5):285-88 (1982). R. B. Held and M. Clawson, *Soil Conservation in Perspective* (Baltimore, Md.: Johns Hopkins University Press, for Resources for the Future, 1965); Hoover and Wiitala, "Operator and Landlord Participation in the Maple Creek Watershed in Northeast Nebraska;" L. K. Lee, "The Impact of Landownership Factors on Soil Conservation," *American Journal of Agricultural Economics* 62(5):1070-1076 (1980).

19. Crittenden, "Soil Erosion Threatens U.S. Farm Output."

20. Lee, "The Impact of Landownership."

21. P. M. Raup, *The Federal Dynamic in Land Use*, Report no. 180 (Washington, D.C.: National Planning Association, July, 1980).

22. R. A. Collins, "Federal Tax Laws and Soil and Water Conservation," *Journal of Soil and Water Conservation* 37(6):321 (1982).

23. *Ibid.*, p. 322.

24. B. Johnson and M. Baker, "The Impact of Tax Policy on Soil Conservation," paper presented at the Annual Meeting of the American Agricultural Economics Association, July, 1980.

25. Held and Clawson, *Soil Conservation in Perspective*; and Hoover and Wiitala, "Operator and Landlord Participation in the Maple Creek Watershed in Northeast Nebraska."

26. Castle and Hoch, "Farm Real Estate Price Components 1920-78."

Chapter 6

Present Soil Conservation Programs

In the early 1920s, most decision makers thought that landowners, once made aware of the seriousness of soil erosion, would conserve without any public assistance because it was in their long-term interest to do so. By the mid-1930s, this belief had lost support: "It was becoming clear that a land use policy resting upon individual choice, rather than public decision making . . . was inadequate."[1] The depression of the 1930s, the drought, and the persuasiveness of Hugh Bennett ultimately culminated in soil conservation legislation. However, the resulting programs did not meet with unanimous support.

When the 1936 Soil Conservation and Domestic Allotment Act gave authority to the Soil Conservation Service (SCS) to assist farmers in combating soil erosion or flooding, the extension personnel at land-grant universities felt threatened because this activity had traditionally been theirs. The resulting antagonism between the SCS and the universities slowed adoption of the conservation programs, with each group seeking to develop support for its own position.

In addition, many federal agencies opposed the SCS. According to R. Burnell Held and Marion Clawson, in *Soil Conservation in Perspective*:

> Many people in the Department of Agriculture considered soil conservation a much less serious problem than did Bennett and favored a much more gradual approach, primarily through education. All this contention was exacerbated by personal rivalries, which were perhaps inevitable; many resented Bennett's efforts, probably the more so because his efforts had been so successful. The other bureaus and agencies of the Department, and Secretary [of Agriculture] Henry A. Wallace made strenuous efforts to curb and delimit the scope of SCS activities. The Cooperative Extension Service, within the Department, but particularly in the colleges, made special effort to prevent SCS from working directly with farmers other than those for whom it was

doing special demonstration work.[2]

Furthermore, passage in 1937 of the Standard State and Soil Conservation District (SCD) model law brought new opposition. The legislation sponsored creation of local governmental units—soil conservation districts—with the potential to exercise a great deal of authority over land-use planning. Many states opposed such authority. In an effort to speed adoption of the model law, President Franklin D. Roosevelt wrote an eloquent plea to each state governor:

> My Dear Governor:
>
> The dust storms and the floods of the last few years have underscored the importance of programs to control soil erosion. I need not emphasize to you the seriousness of the problem and the desirability of taking effective action, as a Nation and in the several States, to conserve the soil as our basic asset. The Nation that destroys its soil destroys itself. . . .
>
> Very sincerely yours,
> Franklin D. Roosevelt

Ultimately, the states did adopt the standard act in various forms. Essentially, all privately held farmland in the United States is now encompassed in one of the 2,950 conservation districts.

At the same time that the model law was being drafted, the Supreme Court had declared the existing Agricultural Adjustment Act unconstitutional. The act had allowed the government to "enter into agreements with farmers to control production by reducing acreages devoted to basic crops, to store crops on the farm and make payment advances on them and to enter into marketing agreements . . . to stabilize product prices and to levy processing taxes."[3]

In an effort to pay farmers to limit production despite the court's ruling, Congress amended the Soil Conservation Act by a new bill, the Soil Conservation and Domestic Allotment Act of 1936. This act enabled farmers to receive soil conservation payments for reducing "soil-depleting" crops, which were also the crops that were in surplus. Two goals—soil conservation and the maintenance of farm income—could thus be pursued simultaneously.

A considerable battle developed over whether the Agricultural Adjustment Administration (AAA) or the SCS would administer the soil conservation payments. With the support of the leaders of the land-grant universities, the decision was made to allow the SCS to administer technical assistance and the AAA to administer the payments. Ultimately, the soil conservation program was made independent of production-control programs and became a cost-sharing program for the voluntary adoption of soil conservation practices. The AAA eventually became the Agricultural Stabilization and Conservation Service (ASCS).

The ASCS and the SCS have competed for power and support through most of their history. Indeed, the battles over who would administer conservation programs grew so fierce and generated so much criticism that in 1951 Secretary of Agriculture Charles F. Brannan issued Memorandum 1278, which delineated the responsibility of each departmental agency for the soil conservation mission. Despite the secretary's effort, conflict continued.

In 1956, Congress established a new conservation program, the Soil Bank. The Soil Bank was first established by the Agricultural Act of 1956 and was essentially a program to control the supply of farm products. Under the program, farmers received federal payments if previously harvested croplands were placed in soil-conserving uses such as pasture. Thus, once again a program combined the goals of soil conserving and maintaining farm income. Approximately 28.7 million acres were placed in the Conservation Reserve, or Soil Bank, between 1956 and 1960. After 1960, the program was no longer available to new entrants. The last of the contracts expired in 1970.[4]

There appeared to be several reasons for the termination of the Soil Bank program. In the mid-1950s whole farms were put into reserve, leading to complaints from agricultural suppliers and community leaders when farmers no longer purchased farm supplies. Also many people were philosophically opposed to paying farmers for not producing. Another criticism was that it was an exceptionally expensive program that had little effect on production since farmers would farm those fields not dedicated to the Soil Bank program more intensively.[5]

During the 1960s, Congress adopted several other programs that included soil conservation goals. Among these programs was the 1965 Appalachian Regional Development Act, which provided assistance to landowners in the Appalachian Mountain region to control erosion and to stabilize and reclaim land. This added a regional approach to an ongoing program of rural development. Another program was the Food and Agricultural Act of 1962, which provided assistance to rural communities to meet economic development goals, including soil conservation.

Major Federal Conservation Programs

At present, there are 3 major federal soil conservation programs, with many ancillary ones.* The 3 major programs are the Conservation Operations Program (COP), authorized by the 1935 Soil Conservation Act; the Great Plains Conservation Program (GPCP), authorized in 1956 to combat the special soil erosion problems of the Great Plains region; and the Agricultural Conservation Program (ACP), authorized as part of the Soil Conservation and Domestic Allotment Act of 1936 to serve as the mechanism by which the federal government would share the cost of conservation practices.

In addition to these 3 major programs, which focus on the problems of maintaining soil productivity and reducing flood hazards, there are smaller programs that have grown out of the legislation for improved air and water quality.

Conservation Operations Program

COP provides funding for the SCS to provide technical assistance to farmers and ranchers for soil conservation measures. The SCS administers the program in cooperation with local soil and water conservation districts. The programs are primarily voluntary. That is, if a farmer desires assistance and contacts the district conservationist, then technical assistance is offered. While farmers are not legally compelled to cooperate with the districts or to participate in the program, they are encouraged to do so through media publicity, peer pressure, and demonstration projects.

Since 1936, the SCS has expended $3.6 billion through COP, mainly to provide technical advice to farmers and to meet personnel requirements for developing farm plans to aid farmers in implementing soil conservation practices.

---

*For example, the U.S. Department of Agriculture (USDA) has 34 programs that are termed conservation programs, including several for watershed protection,[6] plus a program of educational extension administered by the Science and Education Administration.

Great Plains Conservation Program

GPCP is similar to COP but offers cost-sharing along with the technical assistance. With this program, the SCS aids Great Plains farmers by entering long-term agreements of 3 to 10 years. The contract specifies the type of conservation practices to be followed and the amount of cost-sharing that will be made available. A farmer who fails to comply with the contract can be forced to repay the monies received. Cost-sharing is limited to $25,000 per contract. The level of cost-sharing cannot exceed 80 percent of the cost of any one practice. As of 1980, over 50,000 farmers had participated in the GPCP. Over 5,000 more were waiting for funds to become available.

Agricultural Conservation Program

The Agricultural Conservation Program (ACP) is administered by the ASCS and provides both long-term (3- to 10-year) and short-term (1-year) agreements for financing soil conservation practices. Under ACP, the federal government will pay for 50 to 75 percent of the cost of approved practices, up to a maximum of $3,500 per farmer per year. A local agricultural stabilization and conservation committee, elected by farmers, recommends which problems and solutions appear to be appropriate for cost-sharing. The SCS provides the technical advice for establishing and implementing the approved practices.

Assessing the Programs

Until recently, the effectiveness of these programs in meeting soil conservation goals was not assessed. For most of the years since the inception of the programs, conservation goals coincided with those of agricultural commodity programs. Because crops were in surplus, farmers were paid to reduce the number of acres devoted to crops. Fewer cropland acres generally meant less erosion.

The goals of conserving soil and maintaining farm income continued to mesh reasonably well until World War II. Then, increased farm production was needed. Federal funds were allocated to production-improving practices to bring about an increase in farm output. Despite the return to surplus production conditions following World War II, cost-sharing for production-oriented practices continued until 1980.

The obvious inconsistency of using conservation programs to provide cost-sharing that would enhance production did not go unnoticed. Every president since Harry S Truman attempted to reduce funding for such programs or to eliminate cost-sharing for production-oriented practices, or both. Congress consistently overrode these attempts.

Yet the major agricultural programs of the 1950s and 1960s—Soil Bank and GPCP—provided financial incentives to remove land from crop production. The lands retired were frequently unproductive and erosion prone. Thus, some soil was conserved during this period, despite the encouragement of production-enhancing practices.

The situation changed in the 1970s. With the increase in farm exports, the goals of soil conservation and improved farm income seemed to be in conflict. As one observer noted, "We've become like a third world country, mining our natural resources in order to pay for our imports."[7]

The soil conservation programs began to be criticized for failing to meet conservation goals. In 1977, the comptroller general of the United States criticized all of the programs in a report to Congress entitled *To Protect Tomorrow's Food Supply, Soil Conservation Needs Priority Attention*.[8] In this report, the SCS was censured for its failure, within COP, to direct assistance toward areas with critical erosion problems and for spending too much time on plans for individuals. "Many of the conservation plans in SCS files were outdated, forgotten by the farmer, or just not carried out or used as a basis for making farm decisions."[9] The report concluded that:

> The Great Plains Conservation Program has not made satisfactory progress in alleviating soil erosion problems on agricultural lands in the Great Plains . . . Two factors contributing to slow progress appear to be (1) lack of incentive (because of high crop prices and other reasons) for farmers to establish grassland or to maintain it after their GPCP contracts expire and (2) insufficient identification of farmers with high priority conservation needs and encouragement of them to use the program.[10]

The report also pointed out that while it was originally anticipated that 95 percent of program funds would be used to establish vegetative cover, as of 1975 only 27 percent had been used for this purpose. The rest was put to such uses as fencing, livestock water facilities, and irrigation systems. Furthermore, the report stated that some land converted into "permanent" vegetative cover had been converted back to crops when the farmers' contracts expired.

Similar criticisms were leveled at the ACP: "Federal financial assistance is not effectively directed toward critically needed soil conservation practices having the highest payoff for reducing erosion."[11] In particular, the comptroller general criticized the use of cost-sharing for practices aimed more at enhancing agricultural productivity than at soil conservation—drainage systems, land leveling, and liming of fields, for example. The comptroller general argued that the returns on these investments were large enough for the farmers to finance the activities themselves. Also criticized was a provision in the appropriations acts for 1976 and 1977 that allowed farmers and county committees wide latitude in deciding which practices should or should not receive ACP cost-sharing.

Harsh as this report was, other critics of the programs have been even more shrill:

> The statistics go on and on and are shocking . . . upwards of 4.8 billion tons of soil are eroded from agricultural areas each year . . . .
>
> USDA platitudes, conservation rhetoric, mere lip service in the face of such appalling losses are even more shocking than the statistics. We have here a situation of a patient suffering from pneumonia, and the doctor responsible for recovery keeps diagnosing the sickness as a common cold.[12]

There were, of course, responses to such criticisms.

First, there are some production-improving practices that result in conservation. Fences that keep cattle away from overgrazed fields reduce erosion. Careful placement of water troughs spreads livestock out more evenly across arid acreages. Drainage of a wet field for cultivation may financially enable a farmer to remove row crops from a steep hill and reduce surface runoff.

Also, many past practices are the result of the pressures under which the SCS and the ASCS operate. Both agencies need the popular support of farmers and thus have considerable incentive to spread program benefits widely. Furthermore, because farmers participate voluntarily, the SCS and ASCS must work with farms and farmers as they find them.

> The need to develop soil conservation programs for a particular farm as currently bounded, and for a farmer, perhaps with serious limitations of managerial ability and capital, has imposed severe technical and socioeconomic strains on planning. The result has been limited "success" even by SCS standards, and still more limited success by

national or social standards.[13]

Another problem arises from the eligibility criteria of the programs. "The county committees have accepted applications on the basis of who had walked in the door," observed one USDA official. "The committees have a certain amount to spend, and they try to spread it around as best they can. If a farmer comes in who meets the eligibility requirements, he'll have a good chance of getting a share of the money. It doesn't matter if his land is eroding at 1 ton or 20 tons a year, because the selection criteria aren't tied to erosion."[14]

Even if program managers had attempted to focus assistance on farms with the greatest erosion problems, the lack of data on the nature and extent of soil erosion would have limited their ability to do so. Furthermore, the SCS never has had the authority to attempt to change other institutional or social factors such as capital gains treatment, leasing arrangements, or loan policies that may be encouraging soil-depleting practices. Its mandate is far more limited.

Policy Changes

In response to the public criticism of soil conservation programs, Congress eventually demanded evidence that dollars spent for soil conservation were paying off in reduced soil losses. In late 1976, for example, an oversight directive from Senators Talmadge and Dole requested that USDA produce evidence that its conservation programs were having an impact.[15] Ultimately, the result was the passage of the Soil and Water Resources Conservation Act of 1977 (RCA).* This legislation was adopted, in part, to provide a more scientific basis for budget and program decisions and thereby reduce the conflict between the legislative and executive branches. The act has raised conservation issues to a higher place on the nation's agenda, in part because of the required appraisal of the nation's soil. New data and analyses developed because of RCA have considerably improved information for decision making, even at this early stage of implementation. The act has also enlarged the conservation issue to include more than soil productivity.

The RCA also requires that, in addition to an appraisal of

---

*For a detailed discussion of the politics that led to the passage and implementation of RCA, see Leman.[16]

nonfederal soil and water resources, USDA must develop programs for "furthering conservation, protection and enhancement of these resources." The development of a program has been arduous and only recently, in the fall of 1982, was a short-term program presented to Congress. Reflecting on this lengthy development, one observer remarked that:

> Examination of the alternative policy tools by which the objectives could be achieved was probably the most threatening element of the whole [RCA] exercise. It was also for that reason the least developed in terms of past organized efforts and enjoyed the least agreement on what ought to be done and how. But this element may have the most effect on future changes. It is in the most need of shoring up . . .[17]

Since that observation was made in 1981 some "shoring up" has occurred, but much of the momentum appears to be missing in the current situation as agencies and their clientele have their attention diverted from long-range program development while they fight for their hoped-for share of fewer federal dollars.

Several other criticisms of the conservation programs have been met with rule changes required by the Agriculture, Rural Development and Related Agencies Appropriation Act of 1979. Cost-sharing for practices identified as adding to production but having little or no conservation or pollution-abatement benefits has been eliminated. Also, recommendations by county committees for cost-sharing expenditures must now be approved by a state committee and the U.S. secretary of agriculture. The state committee is to give particular attention to projects that conserve water or improve water quality. Funds for cost-sharing "shall be directed to the accomplishment of the most enduring benefits attainable."[18]

Water Quality Programs

Federal water quality programs can also influence a farmer's adoption of soil conservation practices.

The 1972 Federal Water Pollution Control Act (FWPCA) amendments were designed to manage water quality problems, with the main emphasis on controlling pollution from point sources. Although there was no specific provision for federal regulation of nonpoint pollution, the law does establish a goal of attaining "fishable and swimmable waters" by 1983. To achieve that objective, planning must "assure adequate control of [all] sources of pollutants." This requirement—found in Section 208—requires identifying significant nonpoint sources of pollution and setting out procedures and methods to control them.[19] Enacting regulations require state and local action. Many states, with the help of the U.S. Environmental Protection Agency (EPA), have already prepared water quality plans and have begun to work with local farmers.

In 1977, Congress passed the Rural Clean Water Program (RCWP), which deals primarily with the problems of agricultural runoff. Twenty projects and $70 million have been approved so far as an experimental phase. The areas chosen were those designated as the worst, or second-worst, agricultural nonpoint source pollution problems in each state considered. Reelfoot Lake, Tennessee, for example, is one RCWP project. Farmers in the project areas are considered for cost-sharing of up to 75 percent of the costs of agreed-upon improvements as part of 3- to 10-year contracts with RCWP. Total maximum payment to a farmer is limited to $50,000.

RCWP is not exclusively a USDA endeavor. The EPA has provided some funds for monitoring from its Clean Lakes Program and has assisted in the selection of RCWP project sites.

Several problems have slowed implementation of the RCWP. Where farmers are already involved with long-term agreements under ACP, they are hesitant to engage in another cost-sharing program until they have finished the previously committed work. Another problem arises because the farmers eligible for cost-sharing under RCWP (as well as under the ACP) are often renters of the land they farm. They have little incentive to become involved with improvements for which they will receive little benefit. Furthermore, as is also the case under ACP, the amount of money available through cost-sharing is often insufficient to induce farmers to make the very expensive changes that will most improve water quality.

Despite these problems, there is a concerted effort to make the program a success.

> All of us should keep in mind that this is an experimental program. If we expect to be ongoing, we must demonstrate to the Congress that, as a group, we can run the program well and that the taxpayers are getting their money's worth.[20]

The EPA is working with the USDA in another project aimed at water quality improvement. The Model Implementaton Program (MIP) is a cooperative effort supporting the installation of conservation practices in critical areas until RCWP is fully implemented. Even though cost-shares for MIP provided by ASCS may be as much as 90 percent, it remains difficult to induce farmers to enter into more expensive agreements, such as those requiring animal waste facilities or major design changes in irrigation systems.[21]

There are a number of other special projects under way as well. Of the ACP allocation to states, $10 million is to be used for special projects approved by the county committees. The North Fork of the Forked Deer River Watershed in Tennessee is such a project. This area has some of the worst soil erosion problems in the nation as a result of the intensive row cropping practices on many of the 570 farms in the watershed. The watershed is characterized by a rolling topography of highly erosive soils. The project will use $2.7 million in cost-sharing funds over 3 years to help farmers install such conservation practices as permanent pastures, terrace systems, and sediment basins. As one farmer commented, "I had tried to do the work myself before this, but I wasn't successful. I really needed the cost-sharing."

ACP funds are also being used for the Colorado Salinity Project and Uintah Salinity Project to improve on-farm irrigation practices and thus reduce the Colorado River's salinity.

Since 1962, the USDA has been authorized to provide financial and technical assistance to control agricultural pollution, including introducing measures to convert lands unsuitable for cultivation to improved uses such as permanent pasture. In addition, the Farmers Home Administration can provide loans to farmers to help finance soil conservation practices.*

Commodity Programs

Several other government programs affect farmers' soil investment decisions, especially commodity programs designed to maintain prices of crops through price supports, target prices, or other mechanisms.

---

*The future of both Section 208 and RCWP, like many other programs, is in doubt in the current political climate of retrenchment and budget austerity. It is doubtful that these programs' funding will increase or even remain stable in such an environment.

> Farm programs of the Federal government have, in various ways, supported and stabilized farm prices and income since 1933 ... These programs were operated in a way that reduced risk and uncertainty for farmers, affected their expectations of future income potential from farm production activities, and influenced their willingness and ability to invest.[22]

While such programs have usually included the requirement that farmers set aside part of their acreage for short-term conservation uses, they have also encouraged farm specialization and intensive farming practices designed to maximize production of grains—and have operated more than once to penalize the conservation-oriented farmers.

Price support programs often reduce the need for farm diversification by reducing the risk of specialization. A main reason for diversification is to be buffered from market fluctuations: a farmer can avoid dependence on the market price of one crop by selling a variety of agricultural products. Price supports remove the incentive to diversify. Coupled with disaster payments, tax advantages, and special laws, the programs have encouraged large-scale, specialized farms.[23]

The trend toward such farms encourages depletive management systems in two ways. First, contouring and terracing create problems for the big machinery typically used on large-scale farms. Second, the lack of a crop rotation system, as is often the case with specialized farms, often results in considerable erosion. Although several studies in the 1960s showed that soil conservation is profitable on mixed livestock and cash crop operations in which crop rotations included forage and pasture, crop farmers are reluctant to add livestock, since they are not set up to handle animals.[24] Moreover, grain price supports increase the average selling price of grain relative to other crops (and to livestock).

Federal subsidies encourage farmers to plant more of the crops that are under subsidy. Grains, the major subsidized crops, are typically highly erosive; this is especially true of corn. Thus, the government, in effect, is placing a negative incentive on soil conservation. In some cases, price support programs have directly penalized soil-conserving farmers. In 1975, for example, ASCS regulations encouraged some farmers to plow land that they had kept in grass under federal conservation programs. Despite protests from conservationists, the ASCS refused to let farmers count their protected grasslands as normal crop acreages.[25]

The 1983 payments-in-kind (PIK) program pays farmers in

crops (corn, wheat, rice, and cotton) if they agree to produce less. While the main goal of the PIK program is to reduce surplus grain holdings, there will also be some impact on the participating farm's soil erosion as well. Specifically, farmers who wish to participate in PIK must withdraw as much as 50 percent of their acreage from production and place these acres in conservation uses, such as hay production or grazing. The PIK program, however, will be self-terminating, ending when the current excessive stocks of grain are eliminated, perhaps as early as the 1984 crop year. The long-run contribution to conservation goals of the PIK program are therefore negligible.

Local Institutions

Most soil conservation programs are administered through local SCDs.* The SCDs are local organizations established under local law to address soil and water conservation problems. All of the statutes that create SCDs are based on the 1937 Standard State and Soil Conservation District law. Almost all states authorize SCDs to study soil conservation problems, develop conservation plans, and provide financial assistance to private landowners with conservation problems.

Some, but not all, states also give considerable regulatory powers to the districts.** The district may have the power to require particular methods of cultivation, such as contouring, or the retirement from cultivation of highly erosive areas. However, such land-use regulations are rarely used. One reason is the requirement in most states that proposed regulation receive a favorable vote from a high percentage of landowners. Landowners in a district are not usually willing to give up property rights by

---

*Soil conservation districts (SCDs) are referred to in some states as soil and water conservation districts (SWCDs) and in others simply as conservation districts.

**Twenty-six states directly grant such authority to soil conservation districts. They are: Alabama, Arkansas, Colorado, Florida, Georgia, Illinois, Kentucky, Louisiana, Maryland, Mississippi, Montana, Nebraska, Nevada, New Jersey, North Carolina, North Dakota, Oregon, South Carolina, Tennessee, Texas, Utah, Vermont, Virginia, West Virginia, Wisconsin, and Wyoming. Other states could be listed as granting regulatory authority to districts through revised erosion control or sediment statutes. Iowa is one such state.

voting for such regulations.

The program of the Vernon County SWCD in Wisconsin is one example, however, of a regulatory approach.* In the late 1960s, in response to a particularly flagrant abuse of good conservation practices, local farmers approached their SWCD supervisors about the possibility of regulating soil conservation behavior. Specifically the farmers were concerned about an out-of-state farm manager who oversaw a nonresident corporate farm. He planted the farm in "wall-to-wall" corn regardless of field slope and, in the process, destroyed many previously installed conservation practices. Also the chemical runoff from the corporate farm appeared to be threatening neighboring down-slope alfalfa and tobacco farms. SWCD supervisors decided to develop a regulatory ordinance that focused on a single area—the town of Sterling— as opposed to the entire county. Sterling "contained a portion of the abused land; many of its citizens had initiated complaints; well over half the farmers in the town were voluntarily complying with SCS-SWCD standards through farm conservation plans; and most of the remaining farmers followed conservation practices without SWCD plans."[27] Despite such unanimity, passage of the ordinance was difficult, requiring the Wisconsin legislature to amend the voting procedures for SWCD ordinances in the enabling legislation to a majority-vote criterion. The ordinance did obtain voters' approval (145 to 142) in November 1976.

> The agricultural land provision covers all land in parcels of 1 acre or more with a slope of 6 percent or more. A landowner is "in compliance" with the ordinance if one of two general conditions [is] met: (a) the land is farmed in contour strips . . ., or (b) the land is managed according to a SWCD conservation plan for that farm. Exceptions are granted in cases where technical assistance . . . or cost-sharing funds are unavailable . . .[28]

Enforcement of the ordinance is restricted to those cases where a specific complaint has been registered; SWCD personnel do not seek out violators. Nevertheless, many observers feel the ordinance is influencing good conservation behavior.

It may well be that the success of the Sterling ordinance is due mainly to the unique characteristics of the area—when erosion problems exist, the impacts are obvious—as well as the histor-

---

*This discussion draws heavily from Barrows and Olsen.[26]

ically strong support for conservation by local farmers. Such ordinances may be difficult to enact where there are not graphic examples of poor conservation behavior or where few farmers manage their own lands according to SCD standards. Still, the Sterling experience indicates local action is possible.

Almost all SCDs, however, rely on a voluntary program with federal cost-sharing instead of enforcement authority. This is normally administered through a Memorandum of Understanding with the USDA. Nine states—Idaho, Illinois, Iowa, Kansas, Minnesota, Montana, Nebraska, Ohio, and Wisconsin—have state cost-sharing funds as well. Because the conservation districts have more voluntary applicants for cost-sharing than they have funds, and because they are reluctant to require farmers to add erosion control practices unless cost-sharing is available, the districts rarely enforce erosion ordinances.

In the last few years, some conservation districts have begun to expand their interests to include the development of state and regional water quality management plans. This has been encouraged by the EPA.

As early as 1975, EPA became interested in using the regulatory powers of the SCDs as a way to manage nonpoint sources of pollution. Subsequently, EPA made a grant to several Montana state agencies to study how EPA water quality management regulations could be used to control sediment statewide. One pilot project, in Montana's Lewis and Clark Conservation District, enacted an ordinance to institute land-use regulations.[29] In this SCD, if a farmer is in violation and does not correct the violation voluntarily, then the SCD may petition the district court to compel compliance. The time set for the farmer to come into compliance, however, may be delayed if cost-sharing funds are not available.

State Programs

States have shown an interest in strengthening the regulatory features of their soil conservation programs. The initiative to do so appears to have come from two sources: (1) the requirements of Section 208 of the FWPCA amendments and (2) efforts to expand SCD authority through a model state act for soil erosion and sediment control that was proposed in 1972.[30]

In September 1977, EPA issued a memorandum interpreting Section 208, which states that regulatory programs will be required, if they are the only practical method, to ensure that controls on nonpoint source pollution will be implemented.

The model act was prepared by a task force from the workshop on soil erosion that met at the National Symposium on State Environmental Legislation held in Washington, D.C., in March 1972. The Council of State Governments included the model act in its 1973 suggested state legislation.

> The model act contains a number of innovative features. It includes a provision requiring the State Conservation Commission to adopt state-wide guidelines and conservation standards; it delegates implementation authority to soil conservation districts; it provides an enforcement mechanism which requires conservation plans for land disturbing activities; and it sets out an inspection and monitoring requirement; and finally, it authorizes the levying of penalties and injunctions and other enforcement measures against non-compliance. Consistent with the theory of not penalizing agricultural operators to the point of putting them out of business, the act treats agricultural activities somewhat differently than other activities. Land disturbing activities caused by agriculture are exempt from the penalty provisions unless the operator receives at least 50 percent cost-sharing funds to implement the farm's conservation plan.[31]

The result of these dual incentives has been increased interest by the states in soil conservation. However, only 4 states—Iowa, Michigan, Pennsylvania, and South Dakota—have statutes for soil erosion and sedimentation control that regulate agricultural land and practices. While some states, such as New York, Ohio, Minnesota, and Illinois appear to regulate agricultural lands, the statutes have no enforcement or penalty provisions. Virginia's erosion and sediment control law excludes "tilling, planting or harvesting of agricultural, horticultural, forest crops." Hawaii grants authority to counties to enact soil erosion control ordinances, but does not give the state any oversight authority. Montana's law covers only those agricultural practices that directly disturb streams.

Of the 4 states with regulatory and penalty provisions, Iowa has the longest history, and its program is frequently cited as being the most successful and best designed. Its focus is on soil conservation to maintain soil productivity. (Pennsylvania's law, in contrast, focuses on water quality.) Iowa's statute was the first of the statewide sediment control statutes to regulate agriculture and has served in part as a model for other states.

The Iowa Statutes

In the mid-1960s, an Iowa Drainage Laws Study Committee was formed to recommend improvements for a 50-year-old drainage law because of the rapidly rising costs for reclaiming drainage structures impaired by sedimentation. The committee decided that drainage, flood control, water pollution, erosion, and soil conservation practices were interrelated and recommended a comprehensive approach. One result was a 1971 statute that made soil erosion control mandatory. In 1973, Iowa added appropriations for state-funded cost-sharing for soil conservation practices and in 1980 increased the state's regulatory powers.

Before 1971, Iowa had relied on voluntary programs. The 1971 act and the 1980 modifications give SCDs the authority to classify land, to establish soil loss limits for different soil classes (T-values), and to require owners to employ soil and water conservation practices. However, the SCDs cannot specify what practices must be employed by the farmer as long as the applicable T-values are achieved. Further, SCDs are restricted from requiring owners or operators of farmland to refrain from fall plowing. The districts are, however, empowered to inspect land for violations of soil loss limits, either on a written and signed complaint from an owner or occupant of land being damaged from sediment or on reasonable grounds of belief that soil erosion is occurring at more than twice the allowable rate.

To date, the SCDs have not filed complaints on their own, in part because the detailed procedure for filing is not yet complete. Even when it is complete, there is apt to be some reluctance to file. One reason is the SCDs' desire to maintain popularity with their farming clientele. In addition, there are enough demands on SCD funds and expertise in voluntary arrangements to keep the resources more than fully employed without aggressively pursuing violators.

In fact, an aggressive commitment to reducing soil erosion to the T-values established for each field would result in huge financial and labor requirements. The estimated annual cost to "adequately treat" Iowa's lands comes to $285.6 million.[32]

The law has been challenged. In 1979, a Woodbury County, Iowa, farmer filed a complaint with the SCD, alleging that soil loss from the neighboring farms of Ortner and Schrank exceeded applicable soil loss limits and that sediment was damaging the complainant's farms. The SCD investigated and found that the soil loss was in excess of the 5-tons-per-acre limit set by the district. The SCD issued an order presenting Ortner and Schrank with 2 alternatives: either to seed the land to permanent pasture or hay, or to terrace it. Ortner and Schrank did neither. The cost of the terracing would have been over $12,000 to Ortner and approximately $1,500 to Schrank. The state would have provided grants for the remaining costs.

The SCD brought court action, as provided by the statute, and opened the case of *Woodbury County Soil Conservation District v. Ortner*.[33] Ortner and Schrank alleged that the law's provisions requiring landowners to act to prevent soil erosion were unconstitutional, both as a taking of private property without just compensation and as an unreasonable and illegal exercise of the state's police power.

The court held that the statute was a valid and constitutional exercise of the state's police power, and the constitutionality was not altered simply because the defendants had to make substantial expenditures. The statute was ruled to be reasonably related to the legislative purpose of soil erosion control, which was a proper exercise of the police power. The court stated that, "The state has a vital interest in protecting its soil as the greatest of its natural resources, and it has a right to do so."[34]*

Despite their strength, Iowa's statutes contain potential for delay. The 1980 act requires, for example, that a conservation folder be provided for each farm in the state by January 1, 1985, or as soon thereafter as funds permit. SCDs cannot cite a farmer for excessive soil erosion unless his conservation folder is complete and notice has been given of violations of soil loss limits for 3 or more consecutive years. If a complaint is filed against a farmer, and if a farmer will not participate in a voluntary 50 percent cost-sharing program, then the SCD can require the owner to put in conservation practices, provided there are cost-sharing funds available for at least 75 percent of the cost of any permanent soil and water conservation practices.**

---

*As this book went to press, there was litigation challenging various provisions of the Iowa statute.

**It would appear profitable under the Iowa law for one farmer to file against another in a "friendly" action so as to qualify each other for the 75 percent cost-sharing rather than the 50 percent available under regular programs. This has not turned out to be the case, however.

Iowa's program also directly addresses only soil losses and not water quality. When recommended T-values are based on soil productivity criteria alone, many water quality problems may be neglected. For example, Iowa's greatest erosion losses occur on deep soils near the Missouri River. The erosion has little effect on soil productivity because of the depth of the soil, but has considerable impact on water quality.

References

1. R. J. Morgan, *Governing Soil Conservation: Thirty Years of New Decentralization* (Baltimore, Md.: Johns Hopkins University Press, for Resources for the Future, 1975) p. 35.

2. R. B. Held and M. Clawson, *Soil Conservation in Perspective*. (Baltimore, Md.: Johns Hopkins University Press, for Resources for the Future, 1965), p. 44–45.

3. W. W. Cochrane, *The Development of American Agriculture—A Historical Analysis* (Minneapolis, Minn.: University of Minnesota Press, 1979, P. 141.

4. R. S. Dallavalle and L. V. Mayer, *Soil Conservation in the United States: the Federal Role Origins, Evolution and Current Status*, CRS Report No. 80-144S (Washington, D.C.: Congressional Research Service, September 1980).

5. R. A. Kramer, "Participation in Farm Commodity Programs with Implications for the Changing Structure of Agriculture," unpublished Ph.D. dissertation (Davis, Calif.: University of California, 1980), p. 19.

6. For more information on watershed protection, see W. D. Rasmussen, "History of Soil Conservation, Institutions and Incentives," in H. G. Halcrow, E. O. Heady, and M. L. Cotner, eds., *Soil Conservation Policies, Institutions and Incentives* (Ankeny, Iowa: Soil Conservation Society of America, 1981), p. 3–19; and B. H. Holmes, *History of Federal Water Resources Programs and Policies, 1961–70*, Miscellaneous Publication no. 1379 (Washington, D.C.: U.S. Department of Agriculture, 1979).

7. Quoted in A. Crittenden, "Soil Erosion Threatens U.S. Farm Output," *New York Times*, October 26, 1980.

8. U.S. Comptroller General, *To Protect Tomorrow's Food Supply—Soil Conservation Needs Priority Attention*, CED 77–30 (Washington, D.C.: General Accounting Office, 1977).

9. *Ibid.* p. 10.

10. *Ibid.* p. 47.

11. *Ibid.* p. 27.

12. T. Barlowe, "Three Quarters of the Conservation Job Not Being Done," *Soil Conservation Policies: An Assessment* (Ankeny: Iowa: Soil Conservation Society of America, 1979), p. 129.

13. Held and Clawson, *Soil Conservation in Perspective*, p. 56.

14. K. E. Cook, "Problems and Prospects for the Agricultural Conservation Program," *Journal of Soil and Water Conservation* 36 (1):26 (1981).

15. L. W. Libby and J. L. Okay, "National Soil and Water Conservation Policy: An Economic Perspective," unpublished paper presented at the an-

nual meeting of the Northeastern Agricultural Economics Council June 18–20, 1979, Newark, Del.

16. C. Leman, "Political Dilemmas in Evaluating and Budgeting Soil Conservation Programs: The RCA Process," in Halcrow, Heady, and Cotner, eds. *Soil Conservation Policies, Institutions and Incentives* p. 47–88.

17. D. J. Allee, "Implementation of RCA: A Problem Accommodating Economics in Soil and Water Conservation," in Halcrow, Heady, and Cotner, eds., *Soil Conservation Policies, Institutions and Incentives*, p. 93–108.

18. 7 C.F.R. sect. 701 (1981).

19. B. H. Holmes, *Institutional Bases for Control of Nonpoint Source Pollution under the Clean Water Act with Emphasis on Agricultural Nonpoint Sources* (Washington, D.C.: Environmental Protection Agency, WH-554, 1971), p. 165.

20. R. Fitzgerald, "Conservation Efforts and the Agricultural Stabilization and Conservation Service," in *Soil Conservation Policies: An Assessment* (Ankeny, Iowa: Soil Conservation Society of America, 1979), p. 85.

21. D. T. Massey, *The Model Implementation Program—A Cooperative Effort by USDA and EPA for Water Quality Management: An Overview*, NRED 80-13 (Washington, D.C.: U.S. Department of Agriculture, 1980).

22. F. J. Nelson and W. W. Cochrane, "Economic Consequences of Federal Farm Commodity Programs 1953–72," *Agricultural Economics Research* 29 (2):52 (1976).

23. P. M. Raup, *The Federal Dynamic in Land Use*, Report no. 180, (Washington, D.C.: National Planning Association), p. 3–15.

24. Held and Clawson, *Soil Conservation in Perspective*.

25. J. Risser, "Soil Erosion Creates a Crisis Down on the Farm" *The Conservation Foundation Letter*, December, 1978, p. 3–4.

26. R. Barrows and C. Olson, "Soil Conservation Policy: Local Action and Federal Alternatives," *Journal of Soil and Water Conservation* 36 (6):312–16 (1981).

27. *Ibid.*, p. 314.

28. *Ibid.*, p. 315.

29. M. M. Garner, "Regulatory Programs for Nonpoint Pollution Control: The Role of Conservation Districts," *Journal of Soil and Water Conservation* 32(5): 199–204 (1977).

30. T. R. Henderson, *Mandatory Regulation of Erosion of Agricultural Lands: State Laws and Implementation* (M.S. Thesis) (Madison, Wis.: University of Wisconsin, 1980).

31. *Ibid.* p. 38.

32. This represents a total present value of $24 billion for 50-year project at a 12 percent discount rate.

33. *Woodbury County Soil Conservation District v. Ortner*, 279 N.W. 2d 276 (Iowa 1979).

34. *Ibid.*, p. 278.

Chapter 7

Strategies For Encouraging Soil Conservation

Soil erosion problems are real and, as discussed in previous chapters, are a product of numerous factors. There are technical solutions for controlling erosion on most croplands, but obstacles block their adoption and implementation. Individual farmers make decisions as to the use of our nation's soil; these decisions are influenced by many incentives, constraints, and attitudes. Ultimately, these farmers must take action on their lands if any change in the status quo is to be achieved.

If the outcome of farmers' decisions influenced only their own welfare, there would be little argument for public concern. But the gamble on the importance of soil in future production technologies is a gamble with the next generation's inheritance. Furthermore, today's soil erosion problems are also water and air pollution problems. True, there is much uncertainty surrounding the importance of soil erosion, but it is precisely this uncertainty that makes the case for public concern and public action: "In the absence of good information we should be cautious with our limited soil resources."[1]

The public is becoming increasingly interested in soil conservation, both to protect the long-run productivity of the land and to prevent water and air quality degradation. A crucial question in designing a public policy, however, is how much soil to save? Since conserving every ton of soil would be too expensive, some erosion is acceptable. And there are at least 3 ways to consider how many public resources should be devoted to soil conservation.

One approach is to determine when conservation is economical for society, even if it is not economical for the individual. This is known as the present-value approach. All of the net costs of soil erosion, whether on-farm or off-farm, in the present or in the future, are added to determine whether or not it is profitable to save soil and, if so, where and when. Society would then act to conserve soil on those acres where the additional benefits from increased crop productivity and improved water and air quality are equal to or greater than the costs associated with erosion.*

Although this approach has much to recommend it in theory, actual and accurate quantification of present or future benefits or costs is nearly impossible. In addition, the present-value approach heavily weighs the preferences of today's decision makers rather than preferences of future generations.[3] This may or may not be injurious to future generations, however. The difficulties of quantification mean that today's decision makers must balance the possible but uncertain returns that will accrue to society from an investment in soil conservation versus alternative investment possibilities. Alternative investments include those that will yield returns to future generations on a more certain basis. For example, spending more dollars for conservation means spending fewer dollars in improving job skills and general education levels. If we knew that soil erosion would result in extremely expensive or irreversible effects on future productivity and environmental quality, then a large conservation investment would be appropriate. If this were not the case, a smaller investment in conservation and a larger one in education might be more beneficial in the long run.

Because of these uncertainties surrounding the present-value approach, many argue for the insurance approach, which is based on the belief that "the majority may consider that some resources for future eventualities should be maintained even when there is no apparent longtime economic justification . . . for such conservation. Conservation in this case may be looked upon as a form of insurance against technical changes which may or may not take place."[4] In short, society should be willing to devote more resources to insuring the maintenance of soil productivity than the farmer is to protecting against the "worst case" of high food and fiber costs resulting from soil degradation. Thus, the insurance view is that a society should be willing to act to retain the option for future soil use.

The third approach is the conservation ethic approach. This

---

*For further discussion, see Eleveld and Halcrow.[2]

tack seriously questions whether or not the present generation has the right to discount the benefits to be received by future generations:

> There is wide agreement that the State should protect the interests of the future in some degree against the effects of our irrational discounting, and of our preference for ourselves over our descendents. The whole movement for "conservation" in the United States is based upon this conviction. It is the clear duty of Government which is the trustee for unborn generations as well as for its present citizens, to watch over, and if need be, by legislative enactment, to defend the exhaustible natural resources of the country from rash and reckless spoilation.[5]

With this approach, soil conservation becomes a moral issue. Soil should be husbanded and given to the next generation in a condition that provides for that generation's welfare. There exists "an ethical foundation for soil and water conservation. It speaks to a collective sense of responsibility to those who come after us. It is not a land ethic, as such, but an ethic of human interrelationship."[6] This ethical principle can be placed on theological foundations:

> The cardinal precept not to eat the seed grain, no matter how hard the time, also holds for our relationship to natural resources in general and to all other living species. These are the seeds of life to future generations.... Destruction of parts of the environment is as suicidal as imposing grave risks on future generations; it is a desecration and an act of irreverence to both God and creation.[7]

The ethical arguments, whatever their foundations, are unanimous in suggesting that the rights of landowners to deplete resources should be superseded by a set of ethical conservation principles. J. P. Bruce suggests 4 such principles:

> 1) The economy and the natural environment should provide for a flow of goods and services sufficient to meet societal needs in perpetuity.
> 2) The fundamental stability and productivity of the biosphere should not be risked.
> 3) The value of goods and services is in the use that is made of them in satisfying wants and needs, not in the goods themselves.
> 4) Environmental degradation reduces the value that people get from their lives and from the use of goods and services.[8]

In a certain sense, all of these approaches, present value, insurance, and ethical, are correct despite the fact that none provides quantitative answers to the question of how much soil should be allowed to erode. All 3 approaches can be interpreted as rationales for society's obligation to protect soil's productivity at least to a "safe minimum standard" whenever the costs of doing so are not unreasonably large.[9]

The policy dilemma is both to protect air and water quality and to preserve the option to use more soil in the future, while at the same time not being overly conservative. Soil conservation is not costless any more than soil erosion is. To spend too much on erosion control or to spend dollars inefficiently means society will forego other valued goods such as schools, roads, or aid to the destitute. This is a difficult balance to achieve, a task made all the more difficult by the many dimensions of developing a soil conservation program.

Strategy Choices

Any soil conservation program involves numerous interacting dimensions. Figure 7.1 displays such dimensions graphically.

The first dimension is goals. What is to be accomplished? In the last 50 years, soil conservation programs have evolved to encompass a wide range of goals—prevention of soil productivity losses, maintenance of farm income, improvement of water quality and flood control, and general protection of the environment. In the last few years, the emphasis has been on maintaining soil productivity and improving water quality while de-emphasizing maintaining farm income. This may change if net farm income remains low, as it was in 1981 and 1982.

Then there is the question of who. Who will administer and enforce the program? This is an important consideration because the administering agency or government chosen will influence the final outcome.

> Any new program, no matter what its legislative language or story, will be molded by the agency that is charged with its administration . . . so the political decision as to which USDA [U.S. Department of Agriculture] agency will run a new conservation program is really a decision on the balance of power between federal, state, and local governments. That is not a decision that will be made lightly, by either politicans or bureaucrats.[10]

If the Soil Conservation Service (SCS) administers a program, for example, it is likely to rely less on local politically appointed committees than will the Agricultural Stabilization and Conservation Service (ASCS). Similarly, if the federal government

STRATEGIES FOR ENCOURAGING SOIL CONSERVATION 113

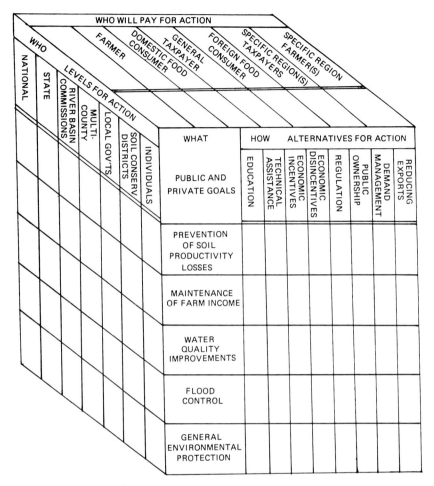

Figure 7.1. Choices for the design of a soil conservation policy

gives block grants to the states to design and administer soil conservation programs, different policies will be developed than if a federal agency administers the program directly.

Critical in the design of any policy is the question: Who will pay? "Who pays is not a side matter to be referred to in an appendix; it is *the issue*."[11] Historically, the voluntary nature of the programs has meant that the general taxpayer paid most of the conservation bills, since farmers are partially compensated for adopting some conservation practices through cost-sharing arrangements. The farmer (and, ultimately, the food consumer and water user) has received most of the benefits.

Programs where costs are widely dispersed and benefits narrowly focused, as is the case with traditional conservation programs, generally are more politically acceptable than when the reverse is true. But acceptance of such policies requires that those affected perceive the policies as equitable. The perception of what is equitable in conservation issues is changing. There is increasing evidence that the owner's right to let land erode will soon no longer be considered either absolute or equitable. The *Woodbury v. Ortner* decision of the Iowa Supreme Court, for example, has given considerable legitimacy to a mandatory approach for achieving soil conservation goals.* Future policies may demand that the farmer share more legal responsibility for soil erosion.

Directly related to the "who will pay" question is the question of how the conservation objectives are to be achieved. There are numerous alternative strategies that can be considered. The most basic is education.

*Education and Technical Assistance***

Education and technical assistance have always been part of conservation programs. They are fundamental in that the decision to add soil conservation practices requires both a perception of the problem and the technical knowledge to do something about it.

Not all farmers can be equally well informed, nor, even when informed, will they see the same need for changes. Also, some farmers who do see the problems and want to add conservation measures may be financially unable to do so. A strategy of education and technical assistance alone therefore is probably of limited effectiveness.***

---

*It is probable that the Iowa Supreme Court was influenced by the existence of cost-sharing arrangements in finding the Iowa statute constitutional.

**The discussion that follows is drawn, in part, from the work of Neely, Carriker, and Rask.[12]

***Education and technical assistance can include such strategies as informing farmers of methods to reduce chemical use without reducing productivity or income. Integrated pest management strategies, for example,

One possible area where education and assistance may make a difference, however, is in the introduction and support of conservation tillage methods. In those regions where the use of no-till or minimum-till improves farm income, programs of education and technical assistance may speed adoption of appropriate practices and thus reduce soil losses. (However, if more acres are farmed with conservation tillage, there may also be reduced water quality because of increased chemical pollution.)

Tax Deductions for Conservation Investments

Existing tax deductions for adopting soil conservation practices could be increased as a strategy to improve conservation. The value of such a deduction would vary with each farmer's personal situation. Farmers with little net income would benefit little from tax relief; farmers in high tax brackets would obtain the most benefits. If, as might be suspected, those farmers with low net income also are those with the most erosive lands, then tax incentives would be ineffective in stimulating conservation on these lands.

Low-Cost Loans

There are several sources of loans for soil conservation practices. The Farmers Home Administration, for instance, will make low-cost loans for pollution abatement and conservation facilities, for erosion control structures, and for establishment of permanent pastures. The Small Business Administration is authorized to finance projects for farmers whose annual receipts are less than $1 million, who would incur substantial hardship without the loan, and who cannot obtain sufficient commercial funds. Additional loans could be provided to encourage more conservation, but they would not be an adequate incentive, unless the investment appears to be profitable.

---

can assist farmers in adopting limited chemical use for pest control. Some organic farming methods can reduce chemical needs without necessarily reducing net income. Both these strategies may improve water quality in some cases as well. Educational and technical assistance programs alone, however, probably cannot produce dramatic reductions in rates of erosion.

Cost-sharing and Direct Payments

Public cost-sharing of conservation practices with farmers has been a major part of traditional conservation programs, and it is likely to continue to be so. Cost-sharing arrangements have the distinct advantage of being easily designated for particular areas and for particular conservation practices. Since many conservation practices are not economical investments for farmers, some form of cost-sharing is probably essential, particularly for investments that will improve water quality but will do little to improve soil productivity. Short-term agreements can give a farmer the needed incentive and financing to implement practices. Long-term agreements help to assure some longevity for any adopted practices.

A variation on the cost-sharing strategy is direct payment to the farmer—in essence, 100 percent cost-sharing, or even greater than 100 percent payments to compensate for any possible reduced yields from adoption of a conservation practice. Another variation is the selection of certain regions and of areas within regions to receive most of the investments. This strategy is known as targeting. Targeting can be a reasonably cost-effective way to reduce erosion, since most sheet and rill erosion occurs on only a small proportion of the nation's total cropland acreage. Almost 70 percent of the nation's erosion exceeding 5 tons per acre per year occurs on less than 8.6 percent of the total acreage.[13]

Despite this concentration of erosion problems, past cost-sharing programs have dispersed assistance widely. Less than 19 percent of the soil conservation practices installed so far have been placed on the most erosive lands. Over half of the cost-shared practices have been placed on lands with erosion rates of less than 5 tons per acre per year.[14] Policies can be made more cost-effective if this pattern were changed and public programs were directed at farmers whose lands account for the lion's share of erosion problems.

Targeting dollars to areas of greater erosion potential, however, should not mean that *every* highly eroding area is protected or reclaimed. Some areas are so severely eroded that thousands of dollars per acre could be spent with little improvement in the land's productivity. The relative benefits to be gained should be balanced against the costs if truly cost-effective strategies are to be developed.

One problem with targeting is establishing criteria for selecting areas. If maintaining soil productivity is the paramount objective, then it may make sense to target those areas with fertile but shallow topsoils, regardless of erosion rates. If water quality goals are primary, then targeting might be focused on areas with high erosion rates, regardless of topsoil depth.

Another issue to be resolved concerning targeting is that of equitable distribution of public assistance. If public subsidies in the form of cost-sharing are given to the farmers with the most severe erosion problems, is past "poor" soil stewardship being rewarded and "good" stewardship slighted? It would appear that, unless carefully designed, the conservation incentives of such a program could be perverse.

Also, there are 2,950 soil conservation districts (SCDs) in the nation—many of which contain croplands that are not eroding severely. Because of this regional diversity, targeting conducted from the national level would have impact on different croplands than if each state received federal funds and directed them at the worst erosion problems within the state boundaries. Similarly, the impact on lands would be different if each SCD received funds and SCD personnel selected the "targets" within their SCD boundaries. Quite understandably, in those districts experiencing less severe erosion, SCD personnel and their farmer clientele feel threatened by a national or state targeting policy. They argue that they have accomplished considerable improvements in an area already and should not be penalized for having less erosive lands or better soil stewards within their district boundaries.

Cross-compliance Strategies

Cross-compliance strategies are incentive programs. The farmer receives extra benefits from other agricultural programs for adopting soil conservation practices, or loses benefits for not adopting them. For example, in what has been termed the "green ticket" approach, farmers might receive higher price support payments for their crops if they had participated in soil conservation programs. In a "red ticket" approach, a farmer who did not participate would lose out on specific federal program benefits.

Cross-compliance strategies have considerable appeal in that they "inject a bit of coherence into federal programs for agriculture. The nation simply cannot live with the situation where one program rewards the farmer for nonconservation behavior

... while another bribes him to conserve."[15]

For cross-compliance to reduce soil erosion effectively, not only would participating farmers need to receive positive net benefits, but the participants would also have to be those farmers whose lands have serious erosion problems. Two of the most likely program candidates for cross-compliance are the price support and acreage diversion programs. Participation in these commodity programs is not evenly spread across the nation, however. Thus cross-compliance linked to these programs would have considerably more impact in some states than in others. Texas, for example, has much of its acreage in corn and cotton, both of which have strong commodity programs. Similarly, Washington has many wheat farmers who elect to participate in commodity programs. West Tennessee, like many other regions in the upper Mississippi Valley, on the other hand, has a substantial amount of land in soybeans, for which few commodity programs exist. Hence, cross-compliance strategies tied to existing commodity programs will favor Texas and Washington over Tennessee and other upper Mississippi Valley states.[16]

To assess the effectiveness of cross-compliance, Dinehart and Libby sampled 400 farmers from 9 wheat and Corn Belt states. They compared the costs of adopting a minimum conservation plan (no planting in straight rows on erosive soils, adding grassed waterways, and using contouring or strip cropping for row crops) to the benefits of participating in price support and production adjustment programs. The benefits included the federal guarantee of a minimum price for wheat or corn. The researchers concluded that:

> 1) cross-compliance will provide a conservation incentive for farmers who are not presently practicing conservation;
> 2) cross-compliance appears to be a cost-effective method of increasing conservation coverage;
> 3) cross-compliance will affect larger farmers, who benefit most from ASCS support programs, more than smaller farmers; [and]
> 4) cross-compliance will increase [the] conservation behavior of those who use land but do not own it.[17]

This last conclusion reflects the fact that commodity programs are usually directed at land operators, while conservation programs are directed at landowners. Cross-compliance could provide the renter-operator of land with an incentive to implement some conservation practices. But the researchers did not estimate the administrative and enforcement costs of such a program.

STRATEGIES FOR ENCOURAGING SOIL CONSERVATION 119

Harvested fields of grain in the Palouse near Colfax, Washington

If cross-compliance is linked to price support or other commodity programs, and if the programs are inactive in times of high agricultural demand, then cross-compliance strategies will not be effective unless long-term benefits are included. For example, farmers might cross-comply to be eligible for *future* program benefits as opposed to *present* ones.[18]

One researcher recommends a variation of cross-compliance: structuring the commodity programs so that only those acres which meet conservation standards would qualify for price supports and other commodity programs. This would give farmers an incentive to expand the number of acres on which soil conservation is practiced and would raise the rental value of such land.[19]

While most discussions of cross-compliance focus on price support and production adjustment programs, there are numerous other programs that could be considered cross-compliance candidates. Loan and credit programs are obvious examples; to obtain a loan at a reduced interest rate, a farmer would have to be in compliance with a soil conservation plan or would have to implement those conservation practices which are relatively low cost, such as retention of residue. Other candidates include disaster and crop insurance programs and state use-value assessment programs, in which farmers are taxed on the agricultural value rather than the market value of their land.

Although it seems likely that cross-compliance strategies, particularly the "red ticket" type, would not be viewed favorably by farmers, a 10-state survey of farmers found that 50 percent either agreed or strongly agreed with the following statement:

> To help achieve national and state soil erosion control goals, each farmer should be required to follow recommended soil conservation measures for his farm to qualify for price and income support programs.[20]

While 39 percent of the farmers were opposed (11 percent had no opinion), 50 percent is still a surprising amount of support for the cross-compliance concept.

Penalties for Soil Loss

Another approach is based on assessing penalties for soil loss. On the basis of the universal soil loss equation (USLE), a farmer could be taxed for each ton of soil lost or paid for each ton of soil saved. The farmers themselves would pay for whatever conservation practices they adopted.

While this strategy appeals because of the direct tie between payments (or taxes) and soil conserved (or soil depleted), the administrative and enforcement costs might prove to be enormous. In addition, accurately measuring soil losses on a per-field basis is a Herculean task. Furthermore, if farmers anticipated a payment for reducing erosion, they would have an incentive to farm, for a time, with the most depletive practices. Even though solutions to these problems might be found, this is not a strategy likely to meet with much favor.

Regulation

While using regulation to achieve soil conservation and water quality goals is certainly a viable option—and one being used in several places—its acceptance is not widespread. Although soil conservation and water quality protection are widely held public objectives, so, too, is the protection of private property rights. As in the case of many resource issues, 2 widely shared objectives are in conflict.

However, the regulatory approach has passed at least one constitutional test (in Iowa), and its acceptability would probably improve if farmers were guaranteed financial assistance in covering the costs of complying with a regulated conservation standard. Perhaps, also, regulations could be directed only to those farmers with the greatest erosion rates. Since minimum tillage, residue retention, and contour-plowing practices can be economical investments in many situations, farmers could be required to adopt such low-cost techniques with little negative effect on income and considerable reduction in soil losses.

With any regulation strategy, however, it is important to recognize that, as increased reductions in erosion are realized, costs tend to increase. In highly eroding areas, it appears that considerable soil savings can be obtained at very little additional cost. However, reductions beyond a certain rate, generally between 2 and 9 tons per acre, are considerably more costly. Rapidly rising costs as soil losses are reduced implies that erosion-rate standards (T-values) must be carefully deliberated. If each dollar spent is to reduce the most erosion possible, then requiring all lands to obtain a specified low erosion rate clearly is not sensible.

Additional Techniques

Changing the tax laws that make speculative purchasing of rural land an attractive investment could encourage some owners to practice more stewardship. Land leases that encourage soil conservation practices and loan officers who require conservation as a condition for borrowing funds could also add to conservation efforts. Other techniques involve increased research. Some research programs will be undertaken by private firms; however, many will need to come from universities. This will require both an infusion of new funds and the assurance of continuity of research funding. Research results that improve the productivity of inputs may reduce the amounts of croplands devoted to the growing of grains and soybeans. For example, developing higher-yielding plants could reduce the amount of land required to meet a given demand; improving the efficiency of the food-processing and -marketing sector to reduce loss and spoilage would be equivalent to increasing yields per acre; developing better weed control for minimum tillage would make conservation tillage techniques more profitable. Other needed research efforts include those designed to reduce uncertainty and to provide information for improving soil conservation policies and encouraging adoption of soil conservation practices. Examples include analyses of the impacts of implementing the various targeting or cross-compliance proposals.*

Other possibilities are more controversial. Reducing food and fiber exports to other nations and improving other countries' capacities for food production with appropriate technology transfers and aid programs would remove some of the pressure from American lands. This might be accomplished with an export tariff or tax. Shifting consumer demand away from grain-fed animals, especially beef, would cut down on acreage used in the more erosive crops of corn and soybeans.

Removing the worst of the highly erosive lands from soybean or grain production is another possibility. These highly eroding lands are often not very productive. For example, 35 percent of all erosion in western Tennessee is estimated to occur on 15 percent of the cropland, which produces only 7 percent of total

---

*See W. E. Larson et al. for a careful discussion of soil conservation research priorities for the nation.[21]

agricultural production in Tennessee.[22]

> The evidence strongly suggests that it is not possible to cultivate inherently erosive land and hold erosion to predetermined levels even by using multiple erosion control practices. Yet this land is responsible for the bulk of the erosion problem. Unfortunately, current programs attempt to reduce erosion to 5 tons per acre while land remains in cultivation—an unrealistic objective that is used to justify bureaucratic programs that consume far too many resources for the amount of soil they save.[23]

Arguing for a removal policy assumes, however, that the benefits to be gained in improving water quality and in using the land for crop alternatives, such as pasture and forest, are considered worth the loss of these lands for crops.

Lands can be removed from crop production by outright purchase or purchased easements, long-term rental arrangements, zoning for types of agricultural use, or mandatory regulations. Each of these suggested policies has drawbacks, not the least of which can be high enforcement costs to ensure farmer compliance.

The Shape of Conservation Policy

The uncertainties surrounding soil conservation choices make the policy choice more one of providing insurance against the worst case possible than of selecting an "optimal" policy. Deciding how much of a premium to pay for this insurance, or how to use that premium in a most cost-effective manner, is not easy. Nevertheless, the worst-case future that might result from inaction is sobering. That vision most certainly demands making some sacrifices today to insure against it becoming a reality.

Implementation of the Soil and Water Resource Conservation Act (RCA) has provided a substantial data base with which to analyze the effectiveness of past soil conservation programs. This data base is augmented by research findings generated or compiled as a result of soil conservation's new visibility on the public agenda. Although there are numerous policy and technical questions yet to be answered, what has been learned to date can serve as a catalyst for genuinely improving soil conservation programs. Where erosion problems occur, what the costs of soil conservation practices are, and how existing programs conflict when achieving conservation goals are all lessons learned and information for future strategy formulations.

In light of these data, some strategies seem to have considerable potential for reducing erosion problems in a cost-effective manner. They include targeting of conservation efforts; removing the most erosive lands from crop production; encouraging or requiring farmers to adopt low-cost conservation practices, such as reduced tillage, residue retention, and contour plowing; and employing some cross-compliance strategies. Also wise insurance investments include funding of research designed to reduce the difficulties of adopting conservation practices (for example, improved weed control with conservation tillage) and of research designed to reduce the need for land through improving yields per acre. Many of these strategies have the advantage of preventing erosion before it begins rather than restoring damages from erosion after the fact.

With 2,950 SCDs, there is considerable chance for experimentation with various strategies. (This is provided that farmers in one area are not unintentionally put at a competitive disadvantage with those from another area because of their participation in an experimental or pilot program.) Because soil erosion problems are as diverse as the nation's topography and people, however, there is reason to be suspicious of any policy so centralized or so inflexible that it cannot be shaped to fit specific circumstances. Despite the advantages of federal coordination, we are apt to achieve greater soil conservation by encouraging a variety of policies rather than by adopting a single national solution. And local SCDs still seem to be a good mechanism for developing and implementing policies, although a role for national programs and leadership remains.

Some of these possible strategies have been encompassed in recent USDA agency planning. In December of 1982, the secretary of agriculture presented to Congress a final *Program Report and Environmental Impact Statement* in response to the provisions of RCA.[24] As part of the final report, the secretary expressed his intent to redirect USDA activities to target 25 percent of SCS and ASCS technical and financial assistance and to consider matching state and local funds by awarding grants to those SCDs experiencing severe erosion problems. The report also details the intent of USDA to request conservation plans from farmers applying for some Farmers Home Administration loans, to emphasize conservation tillage, to resolve inconsistencies in various agency programs, to increase the use of long-term agreements with farmers, and to set up pilot projects to test new approaches

STRATEGIES FOR ENCOURAGING SOIL CONSERVATION   125

for dealing with soil erosion problems. Thus it appears that there will be some redirection of current programs to yield more soil retention or improved water quality per conservation dollar spent.

In the near future, though, increased funding for conservation programs is not in the cards, and some of the changes proposed by the secretary of USDA will encounter considerable opposition.*

There are many political and financial constraints to the adoption of any new conservation programs—vested interests in old programs, limited budgets, limited personnel, traditional views of property rights, and conflicts with other policy objectives. Any forthcoming policy changes, whether at state or federal levels, are therefore likely to be incremental. That does not mean they will be trivial, however.

While the political climate in which new policies will be fashioned is in flux, it is not the climate of a few years or a few decades ago. Recent polls reflect broad public awareness of soil erosion and a willingness to support conservation efforts.[26] Also, the rural population is no longer mainly farmers; farm residents comprise only 15 percent of the total rural population,[27] and there has been a recent revival of rural population growth—almost all of which is nonagricultural. Many of these new residents relocated to enjoy rural amenities.** They are sensitive to environmental quality and recreational opportunities, and are therefore more aware of the off-site costs of soil erosion. Because these new residents feel they have participated in a "back-to-the-land" movement (and because it is not usually their lands which are eroding), it is reasonable to assume that these residents might also be less supportive of the concept that the right to let land erode is not an absolute property ownership right. Also some environmental organizations have decided to make croplands protection a priority issue. The lessons of organizing for political

---

*According to one authority, the excellent evaluation of the Agricultural Conservation Program (ACP) provided by ASCS (which showed both weaknesses and strengths of the program) has been used as an excuse to cut funding to the program, even though the evaluation showed ACP had positive benefits.[25] Such administrative actions give clear signals to agencies not to develop unbiased evaluations of programs in the future.

**For further discussion of this important phenomenon, see Healy and Short.[28]

effectiveness learned through the last 2 decades by these same environmentalists prepared them well to demand placement of conservation on the national agenda as a priority issue.

And while many, perhaps most, farmers perceive themselves to be stewards of the land, there is still a large discrepancy between attitudes and behavior.[29] The challenge is to translate the strong societal desires to avoid scarcity and to maintain a quality environment into laws and other institutional changes that will motivate the farmer of croplands to conserve our nation's soil when and where it seems appropriate.

Little will be accomplished—incrementally or otherwise—without an effective broad-based constituency of persons and groups who are willing to become involved, to participate in political processes in local, state, and federal arenas, and to campaign for soil conservation as a public issue. Mustering this constituency is a particularly difficult task when budgetary austerity is required, the nation has grain surpluses, and farmers have severe credit and cash-flow problems. Nevertheless, conservation is too important to be practiced only when it is convenient.

The evidence is persuasive that the soil erosion problem is real, and in some areas severe. The extent of the problem warrants discussion, examination, analysis, and development of a defensible and cost-effective policy for conserving the nation's soil and water resources.

References

1. L. W. Libby, "A Perspective that Strong Public Action Is Needed to Deal with the Problems of Soil Erosion," paper presented at the *Perspectives on the Vulnerability of U.S. Agriculture to Soil Erosion* symposium of the American Agricultural Economics Association annual meeting, Logan, Utah, August 1-4, 1982.

2. B. Eleveld and H. G. Halcrow, "How Much Soil Conservation is Optimal for Society?," in H. G. Halcrow, E. O. Heady, and M. L. Cotner, eds., *Soil Conservation Policies Institutions and Incentives* (Ankeny, Iowa: Soil Conservation Society of America, 1982).

3. T. Page, *Conservation and Economic Efficiency: An Approach to Materials Policy.* (Baltimore, Md.: Johns Hopkins University Press, for Resources for the Future, 1977).

4. A. C. Bunce, *Economics of Soil Conservation* (Ames, Iowa: Iowa State College Press, 1945).

5. A. C. Pigou, *The Economics of Welfare* as referenced by Bunce, *Economics of Soil Conservation*, p. 97.

6. L. W. Libby, "Economics and Social Realities of Soil and Water Con-

servation," in *Resource Constrained Economies: The North American Dilemma* (Ankeny, Iowa: Soil Conservation Society of America, 1980), p. 155.

7. Science Action Coalition with A. J. Fritisch, *Environmental Ethics: Choices for Concerned Citizens* (New York, N.Y.: Anchor Books, 1980), p. 241.

8. J. P. Bruce, "Choices in Resource Use: Ethical Perspectives," in W. E. Jeske, ed., *Economics, Ethics, Ecology—Roots of Productive Conservation* (Ankeny, Iowa: Soil Conservation Society, 1980), p. 15.

9. C. V. Ciriacy-Wantrup, *Resource Conservation: Economics and Policies* (Berkeley, Calif.: University of California Press, 1982).

10. R. N. Sampson, *Farmland or Wasteland: A Time to Choose* (Emmaus, Pa.: Rodale Press, 1981), p. 289-90.

11. L. W. Libby, "Who Should Pay for Soil Conservation?" *Journal of Soil and Water Conservation* 35(4):155-57 (1980).

12. W. V. Neely, R. R. Carriker, and N. Rask, "Natural Resources: Implications for Agricultural, Environmental and Energy Policies," in *Food and Agricultural Policy Issues for the 1980's*, (Fargo, N. Dak.: North Dakota State University, 1980).

13. C. Ogg and A. Miller, "Minimizing Erosion on Cultivated Land: Concentration of Erosion Problems and the Effectiveness of Conservation Practices," *Policy Research Notes* (Washington, D.C.: U.S. Department of Agriculture, 1981) p. 6.

14. U.S. Department of Agriculture, Agricultural Stabilization and Conservation Service, *National Summary Evaluation of the Agricultural Conservation Program Phase I* (Washington, D.C.: U.S. Department of Agriculture, 1981).

15. L. W. Libby, "Who Should Pay For Soil Conservation?" p. 157.

16. K. Clayton and C. Ogg, "Soil Conservation Under More Integrated Farm Programs" (Unpublished Paper) (Washington, D.C.: U.S. Department of Agriculture, 1982).

17. S. Dinehart and L. W. Libby, "Cross-Compliance: Will It Work, Who Pays?" in W. E. Jeske, ed., *Economics, Ethics, Ecology—Roots of Productive Conservation* (Ankeny, Iowa: Soil Conservation Society, 1980), p. 413.

18. C. Benbrook, "Integrating Soil Conservation and Commodity Programs: A Policy Proposal," *Journal of Soil and Water Conservation* 34(4):160-67 (1979).

19. J. R. Groenewegen, *National Resource Conservation and Agricultural Commodities Programs: A Cross-Compliance Approach* (Draft Manuscript) (Minneapolis, Minn.: University of Minnesota, 1980).

20. H. D. Guither, *How Farmers View Agricultural and Food Policy Issues* (Draft Manuscript) (Champaign, Ill.: University of Illinois, 1981).

21. W. E. Larson, L. M. Walsh, B. A. Stewart, and D. H. Boelter, *Soil and Water Resources: Research Priorities for the Nation* (Madison, Wis.: Soil Science Society of America, 1981).

22. R. Parkins, ed., "Background: West Tennessee Erosion Problem," in *Operation SOS* (Gibson City, Tenn.: SOS Dyer Facelift Committee, 1979).

23. A. Miller, "Approaching Soil Conservation as Though Resources Mattered," a paper prepared for the Midwestern Conference on Food, Agriculture

and Public Policy, November 9, 1981, South Sioux City, Nebr.

24. U.S. Department of Agriculture, *A National Program for Soil and Water Conservation: 1982 Final Program Report and Environmental Impact Statement* (Washington, D.C.: U.S. Government Printing Office, 1982).

25. J. W. Giltmier, "What Priority Conservation?" *Journal of Soil and Water Conservation* 37(5):250-51 (1982).

26. Louis Harris and Associates, *Poll on Rural Environmental Resources Conducted for the Soil Conservation Service* (Washington, D.C.: U.S. Department of Agriculture, 1980).

27. D. A. Brown, "Farm Structure and the Rural Community," ESCSAE Report no. 438, in *Structure Issues of American Agriculture* (Washington, D.C.: U.S. Department of Agriculture, 1979), p. 284.

28. R. G. Healy and J. L. Short, *The Market for Rural Lands: Trends, Issues, Policies* (Washington, D.C.: The Conservation Foundation, 1981).

29. T. L. Napier and D. L. Forster, "Farm Attitudes and Behavior Associated with Soil Erosion Control," in Halcrow et al., eds., *Soil Conservation Policies*, p. 137-50.

Appendix

Measuring Soil Erosion Losses

The first attempt to quantify water-caused erosion was in 1914 by M. F. Mill, a researcher at the University of Missouri. Other institutions and researchers soon joined the effort. After years of experimentation, a cooperating team of scientists from the U.S. Department of Agriculture (USDA) and Purdue University, led by W. H. Wischmeier, developed the universal soil loss equation (USLE). The USLE is specifically designed to measure water-induced rill-and-sheet erosion. The equation was designed to estimate soil loss from fields in the northeastern, southern, and middle regions of the United States. It incorporates 6 factors, all of which can be measured from available data with on-site tests.

The equation is: RKLSCP = A, where A is the average soil loss from a field, usually expressed in tons per acre per year. A is the rate of soil erosion, a product of the following 6 factors:

1) R is the rainfall/runoff factor. A value for R is based on the amount of rainfall and the rate of runoff due to the intensity and duration of rainstorms. R is calculated as the average annual value of the rainfall erosion index (EI). The EI is the product of 2 rainfall characteristics—the total kinetic energy times the maximum 30-minute intensity of the storm.

2) K is the soil erodibility factor. Different soil types have varying susceptibilities to erosion. Values of K include the percentage of silt and very fine sand, the percentage of organic matter, and assigned values for both soil structure (coarseness) and permeability. Because the interest is in the soil's resistance to being moved by erosive forces, K is expressed as the rate of erosion per unit of the EI, for a plot 72.6 feet in length with a 9 percent slope, tilled up and down in continuous fallow.

3) L is the slope length factor. This factor is a ratio of the field's soil loss along the slope length to that of a 72.6-foot slope

under the same conditions. It accounts for the phenomenon that soil loss per unit generally increases as slope length increases. As more water accumulates on a long slope, it has the power to erode and transport more sediment.

4) S is the slope steepness factor. This is a ratio of the field's soil loss to that of a 9 percent slope under the same conditions. Increases in slope mean significant increases in soil loss unless crop cover such as pasture offsets the slope effect.

5) C is the crop cover and management factor. This factor is a ratio of the soil loss from a field of certain cropping management practices compared with an identical area clean-tilled in continuous fallow. The value of this factor accounts for cropping sequence, time between canopies, presence of crop residue, and surface roughness, and this factor differs regionally according to the timing of rains with seasonal harvest.

6) P is the farming practice factor. It is the ratio of soil loss on a field with certain tillage practices to soil loss under straight row plowing up-and-down the slope. Cropland practices to control erosion include contour tilling, strip cropping on the contour, and terracing. On-the-farm water channels to catch excess rainfall can also be part of farming practices.

Past precipitation records and other research data are assembled to determine values for these 6 factors for various areas around the country.

The soil loss prediction equation is meant to serve as a guide for the selection of farming practices. That is, by selecting the allowable level of average soil loss A, and the appropriate farm values for K, L, S, then the cropping and tillage practices (C and P) which equate the relationship can be selected.

There is a modification of the USLE (MUSLE) developed by C. A. Onstad and G. R. Foster that attempts to improve the USLE. Whereas the USLE measures soil dislodged, the MUSLE measures the movement of the soil based on both the amount of energy generated by falling water and the amount of soil that can be dislodged by water striking soil. The MUSLE formula is quite complicated:

$$Y = [0.646 \text{ EI} + 0.45(Q) (q\hat{p})^{0.333} (K) (CE) (PE) (LS)]$$

where:

Y is the sediment yield measured in tons per hectare;

EI is the rainfall energy factor in metric units;

Q is the water runoff in milliliters;
q$\hat{p}$ is the peak runoff rate in milliliters per hectare;
K is the soil erodibility factor;
CE is the crop management factor;
PE is the farming practice factor; and
LS is the slope length and steepness factor

An attempt to quantify soil loss from wind erosion has resulted in a similar equation to the USLE: E = IKCLV, where E is the potential average soil loss due to wind erosion in tons per acre per year.

1) I is the erodibility factor. It is the inherent erodibility of a particular soil and is based on the percentage of the soil particles greater than 0.84 mm in diameter. Larger particles are more stable against breakdown and transport by wind erosion.
2) K is the ridge roughness factor. A standard ratio of ridge height to spacing is 1:4. A comparison of this standard to an actual field-measured ratio can be abstracted to a value for K. Unridged surfaces are more susceptible to wind erosion, provided that ridged rows are oriented at right angles to the prevailing wind direction.
3) C is the climatic factor. It includes average wind velocity and surface soil moisture and temperature measurements.
4) L is the field width factor. It is the measure of the unsheltered distance across a field in the direction of prevailing wind.
5) V is the vegetative cover factor. Its value depends on the kind, quantity, and orientation of crop cover.

The wind erosion equation is a method of estimating wind-induced soil loss so that the various factors can be considered when determining treatment for wind erosion problems.

Index

Agricultural Conservation Program (ACP), 92, 93, 95, 98, 99
Agricultural Adjustment Act, 90
Agricultural Adjustment Administration (AAA), 90–91
Agricultural export market, 5–7, 122
Agricultural productivity, 37–44
Agricultural Stabilization and Conservation Service (ASCS), 81, 91, 93, 95, 99, 112, 124
Agriculture Appropriations Act of 1933 (P.L. 70-769), 4
Agriculture Appropriations Committee, 4
Agriculture, Rural Development and Related Agencies Appropriation Act (1979), 97
Agriculture, U.S. Department of (USDA), 2, 8, 10, 28, 43, 80, 96, 97, 98, 99, 103, 124–25
Air quality and erosion, 44
Appalachian Regional Development Act (1965), 91
Army Corps of Engineers, U.S., 48, 51
ASCS. *See* Agricultural Stabilization and Conservation Service

Bennett, Hugh Hammond, 2, 4, 11, 14–15, 89
Berglund, Bob, 83
"Black blizzards," 17
Brannan, Charles F., 91
Bruce, J. P., 111
Buchanan, James P., 4
Butz, Earl, 5

Chapline, W. R., 2
Chemicals, agricultural, 45–47; *see also* Conservation tillage techniques

Chemistry and Soils, Bureau of (USDA), 2
Chisel plowing. *See* Conservation tillage techniques
Civilian Conservation Corps, 4
Civil War, 2
Clawson, Marion, 89
Clean Water Act, 8
Clean Water Program, Rural (RCWP), 8, 98, 99
Climate and erosion, 18–22
Commodity programs, 99–101
Conservation, benefits of, 75–80
Conservation, costs of, 75–80
Conservation ethic policy approach, 110–11
Conservation, as moral issue, 111
Conservation Operations Program (COP), 92, 94
Conservation policies, approaches to, 109–12
Conservation policies, changes in, 96–97
Conservation policies, future, 123–26
Conservation Reserve. *See* Soil Bank program
Conservation tillage, early use of in U.S., 1–5
Conservation tillage techniques, 60–69; *see also* Soil conservation practices
Conservation tillage trends, 69–71
Construction methods, 59–60
Contour plowing. *See* Conservation tillage techniques
Conventional tillage, defined, 60–62
Cost-sharing, 116–17
Crop removal, 122–23
Crop residue. *See* Residue retention
Crop rotation, 56–57
Crop yields and erosion, 37–44

Cross-compliance strategies, 117–20
Crosson, Pierre, 29, 71

Dinehart, S., 118
Disincentive programs. *See* Penalties
Dole, Robert, 96
Drainage paths. *See* Construction methods

Economic Recovery Act (1981), 84
Education, 114–15
Environmental impacts, 7–8, 44–52
Environmental Legislation, State, National Symposium on, 104
Environmental Protection Agency, U.S. (EPA), 98–99, 103
Erosion, appraisal of, 10–11
Erosion control programs, effectiveness of, 8–10; *see also* Soil conservation programs
Erosion, costs of, 39, 79–80
Erosion and crop yields, 37–44
Erosion estimates, 30–34
Erosion, factors influencing, 18–26
Erosion, measurement tools for, 26–30
Erosion, physical process of, 14–18
Erosion, projections of, 33–34
Erosion, types of, 15–18
Exports. *See* Agricultural export market

Farmers, cost-sharing for, 116–17
Farmers, direct payments for, 116–17
Farmers, education of, 114–15
Farmers Home Administration, 85, 115, 124
Farmers, income of, 77–80
Farmers, as investors, 80–82
Farmers, loans for, 85, 115
Farmers, as owners, 80–82
Farmers, personal preferences of, 74–75
Farmers, planting options, 55–56
Farmers, property rights of, 82–83
Farmers, targeting for, 116–17
Farmers, tax policies for, 83–85, 115
Farmers, technical assistance for, 114–15
Farmers, as tenants, 82–83
Farming practices and erosion, 26
Farmland. *See* Land
Farmland or Wasteland: A Time to Choose, 29
Food and Agricultural Act (1962), 91
Forest Service, U.S., 2
FWPCA. *See* Water Pollution Control Act, Federal

General Accounting Office (GAO), 10
Great Depression, 4, 89
Great Plains Conservation Program (GPCP), 92, 93, 94

Held, R. Burnell, 89
H.R. 7054, 4, 90, 92, 93

Incentive programs. *See* Cross-compliance strategies
Iowa statutes, 105–07
Insurance policy approach, 110
Interior, U.S. Department of the, 4
Investments, tax deductions for conservation, 115

Land, investment in, 80–82
Land, leasing of, 83
Landownership, 80–83
Leasing, changes in, 122
Libby, L. W., 118
Listing. *See* Conservation tillage techniques
Loan policies, 85, 115

McLendon, Bill, 14–15
Measurement tools, 26–30, 129–31
Memorandum 1278, 91
Memorandum of Understanding, 103
Minerals, dissolved, 47
Minimum-till. *See* Conservation tillage techniques

INDEX 135

Model Implementation Program, 98–99
MUSLE. *See* Measurement tools

National Recovery Act of 1933 (P.L. 73–67), 4
National Resources Inventory (1977), 10–11
No-till. *See* Conservation tillage techniques

Payments, direct, 116–17
Payments-in-kind program. *See* Commodity programs
Penalties, 120–21
P.L. 70-769, 4
P.L. 73-67, 4
Plow-plant. *See* Conservation tillage techniques
Policy. *See* Soil conservation policies; *see also* Policy development
Policy development, 123–26
Pollution. *See* Environmental impacts
Present-value policy approach, 109–10
Price support programs. *See* Commodity programs
Program Report and Environmental Impact Statement, 124

RCA, 10–11, 96–97, 123
RCA appraisals, 30, 31–33
RCWP. *See* Clean Water Program, Rural
Reelfoot Lake, Tennessee, 51–52, 98
Regulation, 121
Removal policy, 122–23
Research, need for, 122
Residue retention, 57–59
Resources for the Future, 71
Ridge-plant. *See* Conservation tillage techniques
Roosevelt, Franklin D., 4, 90
Ruffin, Edmund, 2
Ruffin's Folly, 2

Sampson, R. Neil, 29
SCS. *See* Soil Conservation Service; *see also* Agriculture, U.S. Department of
Section 208 (FWPCA), 8, 97–98, 103–04
Sediment basins. *See* Construction methods
Sedimentation, 47–52
Siltation. *See* Sedimentation
Small Business Administration, 115
Soil Bank program, 91, 94
Soil characteristics, alterations of, 55
Soil Conservation Act (H.R. 7054), 4, 90, 92, 93
Soil Conservation District Model Law, Standard State and, 5, 90, 101
Soil Conservation districts, 90, 101–03, 124; *see also* Soil Conservation District Model Law, Standard State and
Soil Conservation and Domestic Allotment Act (1936), 4–5, 89, 90, 92
Soil Conservation in Perspective, 89–90
Soil conservation policies. *See* Conservation policies; *see also* Policy development
Soil Conservation practices, 55–69; *see also* Conservation tillage techniques
Soil conservation programs, administration of, 90–91, 101–03
Soil conservation programs, ancillary, 97–101
Soil conservation programs, assessment of, 93–96
Soil conservation programs, federal, 92–93
Soil conservation programs, state, 103–07
Soil Conservation Service (USDA, SCS), 4, 8, 10, 23, 55, 73, 74, 81, 89–91, 93, 94–95, 96, 112, 124
Soil erosion. *See* Erosion; *see also* Wind erosion
Soil Erosion Service, 4
Soil Erosion: A National Menace, 2
Soil loss, penalties for, 120–21
Soil, properties of, 13–14

Soil type and erosion, 22–26
Soil and Water Resource Conservation Act (P.L. 95-192). *See* RCA
Standards. *See* Soil conservation districts
State Governments, Council of, 104
Strip cropping, 57
Strip-tillage. *See* Conservation tillage techniques
Subsidies, federal. *See* Commodity programs
Sweep-tillage. *See* Conservation tillage techniques

Talmadge, Herman, 96
Targeting, 116–17
Tax policies, 83–85, 115
Tax policies, changes in, 122
Tax Reform Act (1976), 84
Technical assistance, 114–15
Terracing. *See* Construction methods
Tillage. *See* Conservation tillage; *see also* Conservation tillage techniques
Tillage techniques. *See* Conservation tillage techniques; *see also* Farming practices and erosion
Till-plant. *See* Conservation tillage techniques
Topography, alteration of, 55
Topography and erosion, 22–26

To Protect Tomorrow's Food Supply, Soil Conservation Needs Priority Attention, 10, 94
Truman, Harry S, 94

USDA. *See* Agriculture, U.S. Department of
USDA-RCA statistical model, 43; *see also* RCA appraisals
USLE. *See* Measurement tools

Vance, Larry, 29

Wasted Soils: How to Prevent and Reclaim Them, 2
Water Assessment, Second National, 10–11
Water Pollution Control Act, Federal (FWPCA), 8, 97–98, 103–04; *see also* Clean Water Act; Section 208
Water quality and erosion, 44–45
Water quality programs, 97–99
Water Resources Council, 10
Waterways, *See* Construction methods
WEE. *See* Measurement tools
Wheel-track. *See* Conservation tillage techniques
Wind erosion, reduction of, 59
Works Progress Administration, 4
World War II, 93

Yields. *See* Crop yields and erosion